The Patrick Moore Practical Astronomy Series

For further volumes:
http://www.springer.com/series/3192

Using Commercial Amateur Astronomical Spectrographs

Jeffrey L. Hopkins

Springer

Jeffrey L. Hopkins
Phoenix, Arizona, USA

ISSN 1431-9756
ISBN 978-3-319-01441-8   ISBN 978-3-319-01442-5 (eBook)
DOI 10.1007/978-3-319-01442-5
Springer Cham Heidelberg New York Dordrecht London

Library of Congress Control Number: 2013947793

© Springer International Publishing Switzerland 2014
This work is subject to copyright. All rights are reserved by the Publisher, whether the whole or part of the material is concerned, specifically the rights of translation, reprinting, reuse of illustrations, recitation, broadcasting, reproduction on microfilms or in any other physical way, and transmission or information storage and retrieval, electronic adaptation, computer software, or by similar or dissimilar methodology now known or hereafter developed. Exempted from this legal reservation are brief excerpts in connection with reviews or scholarly analysis or material supplied specifically for the purpose of being entered and executed on a computer system, for exclusive use by the purchaser of the work. Duplication of this publication or parts thereof is permitted only under the provisions of the Copyright Law of the Publisher's location, in its current version, and permission for use must always be obtained from Springer. Permissions for use may be obtained through RightsLink at the Copyright Clearance Center. Violations are liable to prosecution under the respective Copyright Law.
The use of general descriptive names, registered names, trademarks, service marks, etc. in this publication does not imply, even in the absence of a specific statement, that such names are exempt from the relevant protective laws and regulations and therefore free for general use.
While the advice and information in this book are believed to be true and accurate at the date of publication, neither the authors nor the editors nor the publisher can accept any legal responsibility for any errors or omissions that may be made. The publisher makes no warranty, express or implied, with respect to the material contained herein.

Springer is part of Springer Science+Business Media (www.springer.com)

Dedication

This book is dedicated to Leah Dawn Hopkins, Kaitlyn Joanne Hopkins, Ryan Lawrence Hopkins and Alexander Jakob Schramm. May my precious grandkids find awe and excitement in science and astronomy for the understanding of the Universe.

Foreword

Jeffrey Hopkins has done it again. Following up on his successful 2012, self-published book, "Small Telescope Astronomical Spectroscopy", he's again gone to a dark place with this new, expanded volume about what is achievable when dispersing light and obtaining spectra with small telescopes.

Having worked with Jeff on two observing campaigns related to the rare eclipses of epsilon Aurigae – in the 1982–1984 time frame, and again for the 2009–2011 eclipse – it has been remarkable to watch as his evolution paralleled the advances in technology over these decades. The key advance is the commercialization of the CCD detector, which has increased light gathering power of telescopes by factors of 1,000. The Nobel Prize committee in 2009 finally recognized the achievement by Willard S. Boyle and George E. Smith "for the invention of an imaging semiconductor circuit – the CCD sensor". Because spectrographs further disperse the sparse incoming stellar photons, photographic spectroscopy had been limited to only large telescopes prior to the CCD revolution.

Hence, in the past decade, affordable CCD cameras plus newly available cost-effective spectrometers, have increased the opportunities for small telescope spectroscopy, and hence the timeliness of this new volume. The role of advanced amateurs in obtaining and analyzing data is becoming ever more crucial in an era when larger telescopes pursue extragalactic science, and/or are being closed due to funding problems. This "citizen science" contribution can sustain research, in ways similar to long-term reports on sunspot numbers, the brightness of variable stars, counts of migrating birds and similar "crowd-sourced" data gathering.

Following a sweeping review of spectroscopy theory, Jeff surveys currently available affordable digital cameras and spectroscopic equipment; he then provides a generic step-by-step tutorial for obtaining spectra. Subsequent chapters explore facets of working with individual instruments including the Star Analyzer, DIY,

ALPY 600 and the LHIRES Spectrographs. This set spans the range of capabilities suitable for small telescopes with finite light gathering capability, from low dispersion, appropriate for exploration of diverse light sources, to moderate dispersion, where quantitative measurement of stellar phenomena becomes feasible. He wraps up with a thorough discussion of available spectrum processing software, particularly Tom Field's RSPEC. The appendices provide handy and useful reference material that will accelerate the user's ability to get out of the dark and into the colorful brightness of astronomical spectra.

If you've read this review, you will like the book!

Robert Stencel
Professor of Astronomy
University of Denver

Preface

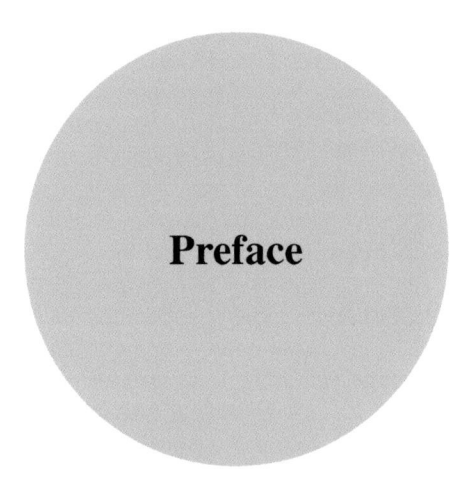

In some astronomical circles, if you are not a professional astronomer, then by default you are an amateur astronomer. What is the difference between a professional astronomer and an amateur astronomer? If a person has a Ph.D. in physics or astronomy, that person is certainly a professional. I have great respect for anyone who has a PhD, particularly in physics or astronomy.

I am called an amateur astronomer, but do professional quality work. I am still labeled an amateur, however. I have written several astronomy-related books, articles for *Sky & Telescope* and a cover feature for *Astronomy* magazine. *Sky & Telescope* even classified me among others as "Super Amateurs." So-called amateur astronomers contribute a great deal to astronomy, but are usually unsung heroes. Oh we hear about an occasional amateur astronomer discovering a new comet or perhaps supernova, but there are actually hundreds of amateurs contributing real science to astronomy every day.

The American Association of Variable Star Observers (AAVSO) is composed mostly of non-professional astronomers. Many of the more serious amateur astronomers have college degrees and many have a PhD in fields other than physics or astronomy. They are all considered amateur astronomers. Everyone associated with astronomy has heard of George Ellery Hale (1868–1938). He was responsible for the giant 40″ Yerkes refractor telescope, Mt. Wilson telescopes such as the Hooker 100″ and the most famous one the Mt. Palomar, the 200″ Hale Telescope. Hale is also responsible for many important discoveries through his observing. Hale had only an undergraduate degree and never received a graduate degree, no Ph.D. Yet Hale was certainly a professional, but by most current definition, he was an amateur astronomer. He surely was not an amateur, however. Clyde Tombaugh (1906–1997), discoverer of Pluto, was an amateur astronomer. He never received a Ph.D., but in most circles he was considered a professional.

While my BS is in physics and I have done graduate study in physics and astronomy, I have never considered myself a professional in those fields. Most of my life's work has been as an electrical engineer. What I find most interesting is that there are also amateur radio operators, HAMs. While they are considered amateurs as opposed to professionals, most of my fellow professional electrical engineers including me were or are also amateur radio operators. We have a HAM license. Many of the advances in radio and electronics resulted from amateur radio operators. I see a great parallel there to amateur astronomers.

One big difference between amateur and many if not most professional astronomers is the amateur usually has more time and freedom to pursue his or her specific interests in astronomy. This is particularly true for those of us who have retired and still possess a desire to make contributions to astronomy. Professional astronomers are usually tied to a university or major observatory. They must struggle to survive. They are limited both in time and as to what projects to pursue. Most amateur astronomers have no such limitations. With your own telescope and observatory in your backyard, you can schedule your projects and observations as you please.

It seems the picture brought to mind when one says amateur astronomer is a person out in a field at night with a telescope looking at stars and other interesting astronomical objects. The person may have trouble finding and identifying astronomical objects other than the Moon. This person is certainly likely to be properly called an amateur astronomer. Perhaps more precisely the term "a beginner amateur astronomer" would be appropriate. More serious observers may participate in what is known as Messier Marathons where during the course of one night all the Messier objects (110 of them consisting of galaxies, nebulae, and star clusters) are observed. While this is of little professional interest, it does show a dedication and mastery of one aspect of astronomy. Others may have a CCD camera connected to the telescope and be taking astronomical images. The telescopes used probably have automatic tracking or may be one of the computer-controlled GOTO telescopes. These amateur astronomers have been taking great images, but there is a great deal more to amateur astronomy than just looking or taking pretty pictures.

Two areas where amateur astronomers have made great contributions over the past 35 years are with astronomical photometry, measuring the changing brightness of stars and asteroids, and within the last 10 years astronomical spectroscopy, obtaining spectra of light from astronomical objects. With the availability of professional-quality, reasonably priced spectrographs, CCD cameras, and telescopes, the field of amateur astronomical spectroscopy is exploding. Just 15 years ago, the idea of doing scientific valuable spectroscopy with a small telescope was considered crazy. One needed very large telescopes and big spectrographs to gather enough photons to produce a good spectrum. When astronomical spectroscopy first was being done, large telescopes and ungainly spectrographs were needed. There were no electronic detectors so photographic plate type cameras had to be used. This complicated things immensely. Plates had to be handled carefully, developed carefully, and then examined in detail manually with a microscope. There were no electronic computers either. Everything was done manually. The revolution started with the availability of high-sensitive CCD detectors and compact spectrographs along with low-cost, high-performance personal computers with corresponding powerful software. These advances have turned the world of astronomical spectroscopy on its head. Advanced amateur astronomers are pushing the envelope with spectroscopy.

The term "amateur astronomer" may need to be revised for the observers doing astronomical photometry and now spectroscopy. We are surely not in the same category as the struggling beginning amateur astronomer out in the field trying to find M31. A possible title for the advanced amateur astronomer is AAA, but that makes one think of cars. Or perhaps we should be called ProAm or PAM. This conflicts a bit with the professional–amateur collaborations, but still fits well. How about SOAs (small observatory astronomers)? An SOA is an advanced astronomer without a PhD, but doing professional-quality astronomy. Or perhaps we should be called PAAs (professional amateur astronomers). A PAA is an advanced amateur astronomer without a PhD. The answer is still not there. Perhaps with the increased advanced amateurs doing serious and professional-quality astronomical spectroscopy, an appropriate name will be forthcoming.

Hopkins Phoenix Observatory　　　　　　　　　　　　　　　　　　　Jeffrey L. Hopkins
Phoenix, AZ, USA

Author's Note

All photographs and figures are by the author unless otherwise noted. All Electronics, RSpec, and Scope Stuff are copyrighted trademarks of their respective businesses. AutoStar Suite, Deep Sky Imager, DSI, DSI Pro, Envisage, LX90, LX200 GPS, and the Meade logo are copyrighted trademarks of Meade Instruments Corporation. Lhires III, LISA, ALPY, and eShel are copyrighted trademarks of Shelyak Instruments. Orion StarShoot and the Orion logo are copyrighted trademarks of Orion Telescopes and Binoculars. ATIK is copyrighted trademark of ATIK Cameras.

Introduction

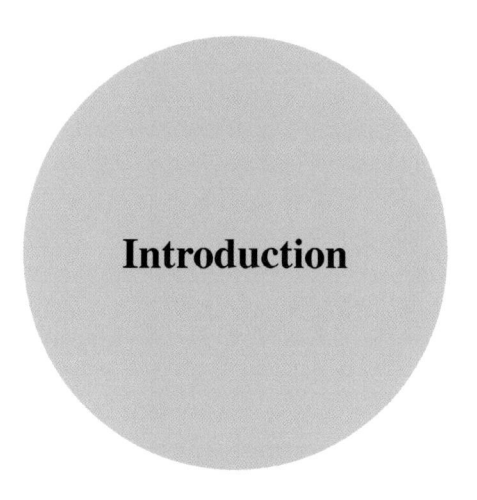

Astronomical spectroscopy is a very exciting type of astronomical observation technique. Once strictly in the domain of large telescope observatories, technical advances have enabled the backyard astronomer to do real astronomical spectroscopy science with reasonably priced and modest equipment. The advanced backyard astronomer with a small to modest telescope, an off-the-shelf spectrograph, laptop computer, CCD camera, sometimes without even a telescope, just a spectrograph and DSLR camera, has the ability to do serious astronomical spectroscopy.

Astronomical spectroscopy is the study of the spectrum of light of astronomical objects, mainly stars. These spectra vary greatly from star to star. What is most interesting is seeing changes in a given star over time. And it is not just stars, but also other astronomical objects such as supernovae, quasars, nebulae, comets, planets and asteroids. Just one spectrum image can contain a wealth of information about an object.

Note: In this Book there will be a lot of talk about wavelengths. In 1866 Swedish physicist Andre Jonas Ångström developed the Ångström unit to describe the wavelength of light. While some low-resolution spectroscopy has wavelengths expressed in nanometers (nm), the wavelengths in this book will be expressed in Ångströms (Å, an uppercase A with a little zero above it). The Å is created on a Macintosh computer by pressing the A – shift and option keys together. For Windows and MS Word select the Å from Symbol under the Insert menu.

One Å = 10^{-8} cm. There are 10 Å in one nanometer (nm).

There are basically two types of astronomical spectroscopy, low-resolution and high-resolution. Low-resolution, which includes mid-range resolution, is where the whole visible spectrum, from below 4,000 Å to over 7,000 Å, is imaged. High-resolution views a small section of the spectrum, typically a spectrum window of 100 Å or less. At a high resolution, more details can be seen and precise measurements made.

This book is designed both as a reference and as an introduction for those observers wishing to know more about what can be done with astronomical spectroscopy as well as those who would like to get started doing spectroscopy with a minimal investment of time and money. There is a detailed discussion of several commercially available astronomical spectrographs available to observers that range from under $200 to close to $5,000. Those prices do not include the telescope or CCD cameras.

The book is divided into eight Chapters.

Chapter 1
Introduction to Spectroscopy
The first part discusses spectroscopy in general with a minimal mathematical introduction and review of the physics of spectroscopy. The second part discusses astronomical spectroscopy. A third part discusses contemporary history of spectroscopy leading to the revolution of amateur astronomical spectroscopy.

Chapter 2
Amateur Astronomical Spectroscopy
This chapter discusses the equipment required, including telescopes, digital cameras, spectrographs, and the basics of taking spectra. A basic knowledge of astrophotography and use of a DSLR or astronomical CCD camera is suggested.

Chapter 3
Star Analyser Low-Resolution Spectroscopy
This chapter discusses low-resolution astronomical spectroscopy with a Star Analyser spectrograph. The intent is to simplify the introduction to spectroscopy. The information presented should allow an observer, using a Star Analyser spectrograph, to be able to obtain a good low-resolution spectrum of a bright "A"-type star, e.g., Vega, with good signal to noise (SNR) and create a wavelength calibrated line profile. Major elemental lines in the spectrum should be easily identifiable. On large red-shifted objects, e.g., Quasars, a measure of the Doppler shift of the hydrogen alpha line and approximate radial velocity and thus red shift should be possible. Other type stars can be imaged and their spectra processed. Even with low-resolution spectroscopy, supernovae can be classified, stars can be classified, large astronomical red shifts measured and elements identified.

Chapter 4
DIY Spectroscopy
This chapter discusses an inexpensive, but professional, Do-It-Yourself (DIY) fiber optic–coupled spectrograph that has been for sale on eBay off and on over the past couple of years. These are refurbished units. A few dozen astronomers have picked up some of these units. This DIY spectrograph is discussed briefly. For the price, if one of these can be obtained, it is an excellent way to experiment with spectroscopy for only $200. The unit is complete with imaging and processing software. The optical bench has a reflective 1,800 l/mm diffraction grating that can be adjusted to cover the whole visible spectrum. Just pointing the fiber optic cable at the outdoors produces a fine spectrum of the Sun. Pointing at various lights and neon lamps shows corresponding spectra. All that is needed is a Windows-based computer.

Introduction	xvii

Chapter 5
ALPY 600 Mid-Resolution Spectroscopy
 This chapter discusses the new ALPY 600 spectrograph. During early 2013, Shelyak introduced a new low to mid-resolution 600 l/mm transmission grating spectrograph called the ALPY 600. While a lot more expensive than the Star Analyser, the ALPY 600 provides a bridge to the higher-end spectrographs. There are three modules that can be purchased separately: the Basic Module, Guiding Module and Calibration Module. The Basic Module is all that is needed to do spectroscopy, but the other modules can add to the usability of the spectrograph. The ALPY 600 provides a bridge between the Star Analyser and the LISA or Lhires III spectrographs.

Chapter 6
Lhires III Mid and High-Resolution Spectroscopy
 This chapter discusses the Lhires III spectrograph. The Lhires III comes with a stock 2,400 l/mm grating. There are optional diffraction gratings available in 150, 300, 600 and 1,200 l/mm resolutions. Two resolutions will be discussed, one for mid-resolution spectroscopy using a 600 l/mm grating and the second for high-resolution spectroscopy using a 2,400 l/mm grating.

Chapter 7
Spectrum Processing Software
 This chapter discusses spectrum-processing software using three different software programs. First is a DIY method using Microsoft's Excel spreadsheet program. While considerably more work than using a dedicated spectrum processing software program, the DIY approach can have good educational benefits by allowing the user to investigate just what is going on with the processing. Two specific spectrum processing programs will also be discussed. One is a French freeware program called VSpec. While free, VSpec can be a large challenge to master and is not what can be considered user friendly for the beginner. The second program is RSpec. While not free, RSpec is highly recommended for its user-friendliness and power. RSpec invites experimentation and thus learning. There is little that RSpec can't do and it can do things much easier than other programs.

Chapter 8
Astronomical Spectroscopy Projects
 This chapter starts with a discussion of various astronomical organizations where amateur astronomers can get involved. Next is a discussion of stellar spectroscopic projects with interesting stars. Additional projects include quasars and supernovae.

Acknowledgments

Dr. Robert E. Stencel (Dr. Bob) for his inspiration and encouragement in my pursuit of astronomy.

Dr. John Martin for his help with understanding some of the finer points of spectroscopy.

Tom Field for his encouragement and development of the RSpec spectrum processing software.

Olivier Thizy for his contribution of professional and affordable spectrographs.

Gene Lucas for his help and encouragement with my astronomical projects.

Pamela Rose for proofreading this manuscript.

Contents

1 Introduction to Spectroscopy 1
   General Spectroscopy Theory 1
   Astronomical Spectroscopy Theory 8
   Modern Astronomical Spectroscopy History 33
   Telescopes 37
   The First Astronomical CCD Cameras 41

2 Amateur Astronomical Spectroscopy 45
   Introduction 45
   Astronomical Spectroscopy Equipment 46
   Spectrographs 55
   Taking Spectra 60

3 Star Analyser Spectroscopy 75
   Introduction 75
   Blazed Gratings 75
   Star Analyser Equipment 77
   Taking the Spectra with a Star Analyser 88
   Image Processing 91
   Low-Resolution Spectrum Processing 92
   Star Analyser Conclusion 99

4 DIY Spectroscopy 101
   Introduction 101
   DIY Spectrometer Specifications 103
   DIY Spectrometer Equipment 104
   Using and Adjusting the DIY Spectrometer 107

	Taking a Spectrum	112
	Spectrum Studio Software	114
	DIY Spectrometer Conclusion	123
5	**ALPY 600 Mid-Resolution Spectroscopy**	**125**
	Introduction	125
	The Module	125
	Basic Module	127
	Guiding Module	133
	Calibration Module	138
	Complete Assembly	139
	Taking the Spectra with an ALPY 600	140
6	**Lhires III Spectroscopy**	**151**
	Introduction	151
	Digital Cameras	151
	The Lhires III Spectrograph	152
	High-Resolution Imaging Technique	162
	Additional Spectrum Processing Considerations	163
	High-Resolution Spectrum Processing	163
	Lhires III Tips	168
7	**Spectrum Processing Software**	**171**
	Introduction	171
	Data Reduction Versus Spectrum Processing	172
	Data Reduction	172
	Spectrum Processing	173
	DIY Spectrum Processing (Excel)	174
	RSpec Spectrum Processing Software	175
	VSpec French Freeware	222
8	**Astronomical Spectroscopy Projects**	**233**
	Introduction	233
	Astronomical Organizations	234
	Stellar Spectroscopy Projects	234

Appendix A	**Astronomical Time**	253
Appendix B	**Fits Header**	263
Appendix C	**Important Wavelengths**	267

Glossary	269
About the Author	279
Index	281

List of Figures

Fig. 1.1	A spectrum produced with a prism	2
Fig. 1.2	Electromagnetic spectrum	3
Fig. 1.3	Hydrogen energy states	4
Fig. 1.4	p/o Hydrogen transition series	5
Fig. 1.5	Hydrogen Balmer spectrum	6
Fig. 1.6	Types of spectra	7
Fig. 1.7	Continuous spectrum	7
Fig. 1.8	Emission spectrum	7
Fig. 1.9	Absorption spectrum	8
Fig. 1.10	Fraunhofer lines in a solar spectrum	9
Fig. 1.11	Wien's displacement law	11
Fig. 1.12	A0V star spectrum	13
Fig. 1.13	B0V star spectrum	14
Fig. 1.14	F0V star spectrum	14
Fig. 1.15	G0V star spectrum	14
Fig. 1.16	G5V star spectrum	15
Fig. 1.17	K0V star spectrum	15
Fig. 1.18	M0V star spectrum	15
Fig. 1.19	RK0V star spectrum	16
Fig. 1.20	WG0V star spectrum	16
Fig. 1.21	Spectrum image versus line profile	18
Fig. 1.22	Selected continuum area for normalization	19
Fig. 1.23	Full width half maximum	20
Fig. 1.24	Equivalent width	21
Fig. 1.25	Equivalent widths and VR of hydrogen alpha	21
Fig. 1.26	Linear – straight line	22

Fig. 1.27	Excel chart type	24
Fig. 1.28	Excel trendline selection	25
Fig. 1.29	Excel polynomial trendline selection	25
Fig. 1.30	Excel polynomial graph	26
Fig. 1.31	Un-calibrated line profile	27
Fig. 1.32	Wavelength calibrated line profile	27
Fig. 1.33	Single point and dispersion wavelength calibration	28
Fig. 1.34	Multiple line/non-linear wavelength calibration	29
Fig. 1.35	Zero and two first order spectra	29
Fig. 1.36	Earth's radial velocity	31
Fig. 1.37	Tangential velocity – 1,042 * latitude	31
Fig. 1.38	Parallax distance measurements	33
Fig. 1.39	Absolute and apparent magnitudes	35
Fig. 1.40	Inverse-square law	35
Fig. 1.41	Cepheid magnitude-period relationship	36
Fig. 1.42	40″ Yerkes refractor telescope	38
Fig. 1.43	Author with Mount Wilson 100″ Hooker telescope	39
Fig. 1.44	Mount Palomar 200″ Hale telescope	40
Fig. 2.1	Hopkins Phoenix observatory	46
Fig. 2.2	Permanent pier and 12″ LX-200 GPS telescope	47
Fig. 2.3	Simple pier for 8″ LX90 and star analyser	48
Fig. 2.4	Wedge and pier	48
Fig. 2.5	8″ LX90 telescope on pier	49
Fig. 2.6	CCD chip pixels	50
Fig. 2.7	Meade DSI Pro series	51
Fig. 2.8	Orion StarShoot G3 camera	52
Fig. 2.9	ATIK 314L+ monochrome CCD camera	52
Fig. 2.10	DSI and Orion Star Shoot CCD spectra response curves	54
Fig. 2.11	ATIK 314L+ ICX285AL CCD spectra response curve	54
Fig. 2.12	Star analyser spectrograph, credit: Shelyak instruments	56
Fig. 2.13	DIY spectrograph	56
Fig. 2.14	ALPY 600 spectrograph, credit Shelyak instruments	57
Fig. 2.15	LISA spectrograph, credit: Shelyak instruments	58
Fig. 2.16	Lhires III spectrograph, credit: Shelyak instruments	59
Fig. 2.17	eShel spectrograph, credit: Shelyak instruments	60
Fig. 2.18	Zero and first order spectra	61
Fig. 2.19	Linearity break determination	63
Fig. 2.20	Hot pixels and dark frame subtraction	64
Fig. 2.21	Pixel map area selection with AutoStar image processing	65
Fig. 2.22	AutoStar suite image processing pixel map	65
Fig. 2.23	RSpec pixel map	66
Fig. 2.24	Orion camera studio pixel count	67
Fig. 2.25	VSpec pixel count	67
Fig. 2.26	Raw flat image	69

List of Figures xxv

Fig. 2.27	Raw flat image pixel map	69
Fig. 2.28	Flat frame pixel map	70
Fig. 2.29	Spectrum and line profile without flat frame correction	71
Fig. 2.30	Spectrum and line profile with flat frame correction	71
Fig. 2.31	Spectrum image orientation	72
Fig. 2.32	Horizontal binning (Bin 2)	73
Fig. 3.1	Non-blazed grating	76
Fig. 3.2	Blazed grating	76
Fig. 3.3	Star analyser	78
Fig. 3.4	Rainbow optics	78
Fig. 3.5	Star analyser and locking ring	80
Fig. 3.6	Star analyser prism/grism	81
Fig. 3.7	Star analyser with a DSLR camera (non-converging mode)	83
Fig. 3.8	Star analyser with an eyepiece	84
Fig. 3.9	Star analyser, locking ring and web cam	84
Fig. 3.10	Star analyser, web cam and star diagonal	85
Fig. 3.11	Star analyser, monochrome CCD camera and star diagonal	85
Fig. 3.12	Zero order spectrum profile F/6.3 (top) versus F/10 (bottom)	86
Fig. 3.13	Expanded spectrum profile F/6.3 (top) versus F/10 (bottom)	87
Fig. 3.14	Bright first order spectrum	89
Fig. 3.15	Solar (G2V) line profile	93
Fig. 3.16	Vega (alpha Lyrae – A0 V) line profile	93
Fig. 3.17	Aldebaran (alpha Tauri – K5 III) line profile	94
Fig. 3.18	Capella (alpha Aurigae – G8 III:+F) line profile	94
Fig. 3.19	Deneb (alpha Cygni – A2 Ia) line profile	95
Fig. 3.20	Betelgeuse (alpha Orionis – M2 Iab) line profile	96
Fig. 3.21	Rigel (beta Orionis B8 Ia) line profile	97
Fig. 3.22	Bellatrix (gamma Orionis – B2 III) line profile	98
Fig. 3.23	Sirius (alpha CMa A – A1 V) line profile	99
Fig. 4.1	DIY spectroscope, front view	101
Fig. 4.2	Sony ILX511 CCD wavelength response	103
Fig. 4.3	Fiber optic cables and connectors	105
Fig. 4.4	Star diagonal detail	106
Fig. 4.5	Relay lens detail	107
Fig. 4.6	Star diagonal and relay lens	107
Fig. 4.7	DIY spectrograph optical bench	109
Fig. 4.8	Mercury spectrum line profile	112
Fig. 4.9	DIY spectrograph for astronomy	113
Fig. 4.10	DIY fiber optic interface	113
Fig. 4.11	Spectrum studio window	114
Fig. 4.12	Spectrum studio window tabs and icons	115
Fig. 4.13	Spectrum studio window icons	115
Fig. 4.14	Spectrum line profile	117

Fig. 4.15	Options – data logging	118
Fig. 4.16	Options – spectral line identification window	119
Fig. 4.17	Spectral line identification	120
Fig. 5.1	In the basic module box	127
Fig. 5.2	Basic module with CCD camera	128
Fig. 5.3	Basic module elements	128
Fig. 5.4	Basic module core element X-ray	129
Fig. 5.5	CCD camera coupler before and after	130
Fig. 5.6	Basic module adjustments	131
Fig. 5.7	Spectrum orientation	132
Fig. 5.8	Guiding module	133
Fig. 5.9	Guiding module elements	134
Fig. 5.10	Coupler with extra C-mount to T-thread adapter	134
Fig. 5.11	Guiding module and core element X-ray	135
Fig. 5.12	Guiding port and reflective slit orientation	136
Fig. 5.13	Guiding port slit focus and orientation	136
Fig. 5.14	Basic module 1.25″, Crayford focuser and 2″ SCT couplers	137
Fig. 5.15	Using the basic module 1.25″ telescope coupler	137
Fig. 5.16	Calibration module, credit Shelyak instruments	138
Fig. 5.17	Calibration module X-ray	138
Fig. 5.18	ALPY 600 spectrograph assembly, credit Shelyak instruments	139
Fig. 5.19	ALPY 600 spectrograph with cameras, credit Shelyak instruments	139
Fig. 5.20	Basic module fluorescent and neon lamp spectra	140
Fig. 5.21	Basic module with neon lamp	141
Fig. 5.22	Filter band pass determination setup	141
Fig. 5.23	Hydrogen alpha filter line profile	142
Fig. 5.24	Basic module on telescope	142
Fig. 5.25	Basic and guiding modules on bench	144
Fig. 5.26	Basic and guiding modules on telescope	144
Fig. 5.27	Computer screen shot of observation	145
Fig. 5.28	Non-linear wavelength calibration (RSpec)	146
Fig. 5.29	Basic and guiding modules spectrum line profile	147
Fig. 5.30	Neon ring calibration	148
Fig. 5.31	Neon ring and low-pressure sodium spectrum	148
Fig. 5.32	Neon calibration	149
Fig. 5.33	Basic module fiber optic coupler	150
Fig. 5.34	Basic module fiber optic telescope interface	150
Fig. 6.1	Lhires III spectrograph X-ray, credit: Shelyak instruments	152
Fig. 6.2	Lhires III spectrograph on the bench	153
Fig. 6.3	Optional 600 l/mm grating unit	154
Fig. 6.4	Neon and laser pointer lines around hydrogen alpha	155
Fig. 6.5	Lhires III with 2,400 l/mm grating micrometer calibration	156
Fig. 6.6	Lhires III with 600 l/mm grating micrometer calibration	156

List of Figures

Fig. 6.7	Lhires III mounted on 12″ LX200 telescope	157
Fig. 6.8	Guide camera slit image	159
Fig. 6.9	Lhires III neon calibration and focusing side plates	159
Fig. 6.10	Lhires III neon line focusing doublet	160
Fig. 6.11	Unfocused neon lines	161
Fig. 6.12	Focused neon lines	162
Fig. 6.13	Spectrum delimiting lines	164
Fig. 6.14	Neon lamp ring wiring	165
Fig. 6.15	Telescope neon lamp ring calibrator	166
Fig. 6.16	18-W low pressure sodium	167
Fig. 6.17	Fulham digital ballast	167
Fig. 6.18	Fulham digital ballast wiring	168
Fig. 6.19	Digital ballast wiring and glowing sodium light	168
Fig. 7.1	DIY data text file and trimmed file	174
Fig. 7.2	Excel line profile of DIY spectrum data	175
Fig. 7.3	RSpec spectrum image and line profile windows	177
Fig. 7.4	Tools menu, option selection	178
Fig. 7.5	Basic program options	178
Fig. 7.6	Advanced program options	179
Fig. 7.7	File loading	180
Fig. 7.8	Spectrum window orange spectrum delimiting lines	181
Fig. 7.9	No visible spectrum	182
Fig. 7.10	Histogram tool	183
Fig. 7.11	Histogram sliders	183
Fig. 7.12	Histogram image enhancement	184
Fig. 7.13	Noisy line profile	184
Fig. 7.14	RSpec pixel map selection	185
Fig. 7.15	RSpec pixel map ADU counts	186
Fig. 7.16	Select rotate option	187
Fig. 7.17	Rotate and slant slides	187
Fig. 7.18	No sky subtraction	188
Fig. 7.19	Subtract background/sky option	189
Fig. 7.20	Sky/background subtracted	189
Fig. 7.21	Selecting horizontal binning	190
Fig. 7.22	Line profile window icons	190
Fig. 7.23	Initial line profile window	191
Fig. 7.24	Measure icon	191
Fig. 7.25	Measurement information options	192
Fig. 7.26	Line profile window vertical measure lines	192
Fig. 7.27	Measure results	193
Fig. 7.28	Additional selections	194
Fig. 7.29	Selecting appearance	194
Fig. 7.30	Labels menu	195
Fig. 7.31	Control of lines in the line profile	196

Fig. 7.32	Points visible	196
Fig. 7.33	Graph labeling the profile title	197
Fig. 7.34	Deselect legend visible	198
Fig. 7.35	Reference menu	198
Fig. 7.36	Edit menu	199
Fig. 7.37	Delete range	199
Fig. 7.38	Full image line profile	200
Fig. 7.39	Trimmed/expanded line profile	200
Fig. 7.40	Delete area between white lines	201
Fig. 7.41	Delete line of line profile	202
Fig. 7.42	Math option	202
Fig. 7.43	Math menu	203
Fig. 7.44	Wavelength-calibration	203
Fig. 7.45	Calibration menu	204
Fig. 7.46	Bubble pixel position	205
Fig. 7.47	Barycenter pixel position	205
Fig. 7.48	Barycenter detail	206
Fig. 7.49	Two point wavelength-calibration	207
Fig. 7.50	One point wavelength-calibration	208
Fig. 7.51	Non-linear window	209
Fig. 7.52	Selecting elemental lines	210
Fig. 7.53	Data file editing	211
Fig. 7.54	Program star line profile trimmed and calibrated	213
Fig. 7.55	Program star line profile smoothed	213
Fig. 7.56	Program star line profile splined	214
Fig. 7.57	Curve A	214
Fig. 7.58	Standard star line profile (A0V.DAT) trimmed	215
Fig. 7.59	Standard star smoothed	215
Fig. 7.60	Standard star line profile splined	216
Fig. 7.61	Curve B	216
Fig. 7.62	Curve A and curve B	217
Fig. 7.63	Math series curve A/curve B	217
Fig. 7.64	Response calibration curve C	218
Fig. 7.65	Program line profile and calibration curve C	218
Fig. 7.66	Response calibrated line profile	219
Fig. 7.67	Synthesize option and control	219
Fig. 7.68	Synthesized low-resolution pseudo spectra	220
Fig. 7.69	Synthesized high-resolution pseudo spectra	220
Fig. 7.70	Hydrogen alpha normalized line profile	221
Fig. 7.71	Equivalent width measurement	222
Fig. 7.72	VSpec opening window select fits file	223
Fig. 7.73	VSpec with spectrum image	224
Fig. 7.74	VSpec spectrum icons	224
Fig. 7.75	Spectrum delimiting	225
Fig. 7.76	Spectrum line profile	226

List of Figures

Fig. 7.77	Graduations icon	226
Fig. 7.78	Selecting multiple lines calibration	227
Fig. 7.79	Selecting elements	228
Fig. 7.80	Wavelength selection	229
Fig. 7.81	Wavelength calibrated profile	229
Fig. 7.82	Heliocentric correction selection	230
Fig. 7.83	Heliocentric correction menu	231
Fig. 7.84	Heliocentric correction results	231
Fig. 8.1	Constellation Aurigae	235
Fig. 8.2	Epsilon Aurigae system diagram	236
Fig. 8.3	Epsilon Aurigae hydrogen alpha line profile	237
Fig. 8.4	Epsilon Aurigae sodium D line profile	238
Fig. 8.5	Orion constellation	239
Fig. 8.6	Betelgeuse low-resolution line profile	240
Fig. 8.7	Betelgeuse mid-resolution line profile	240
Fig. 8.8	Delta orionis low-resolution line profile	241
Fig. 8.9	Delta orionis mid-resolution line profile	242
Fig. 8.10	Delta orionis high-resolution line profile	242
Fig. 8.11	Epsilon orionis mid-resolution line profile	243
Fig. 8.12	Constellation scorpius	244
Fig. 8.13	Constellation cygnus	246
Fig. 8.14	P Cygni typical spectrum line profile	247
Fig. 8.15	Sample WR star spectrum line profile	247
Fig. 8.16	Be star hydrogen alpha emission lines credit: Shelyak	249

List of Tables

Table 1.1	Hydrogen series	5
Table 1.2	MK of spectral type	12
Table 1.3	Luminosity classes	13
Table 1.4	Polynomial pixel versus wavelength table	24
Table 2.1	CCD chip specifications	53
Table 4.1	Available gratings	102
Table 4.2	Element identification	120
Table 5.1	Spectrum window versus CCD pixel size	130
Table 6.1	Laser pointer wavelengths	156
Table 6.2	Neon line pixel x-axis position with internal neon calibrator	165
Table 7.1	Select element wavelengths	211

Chapter 1

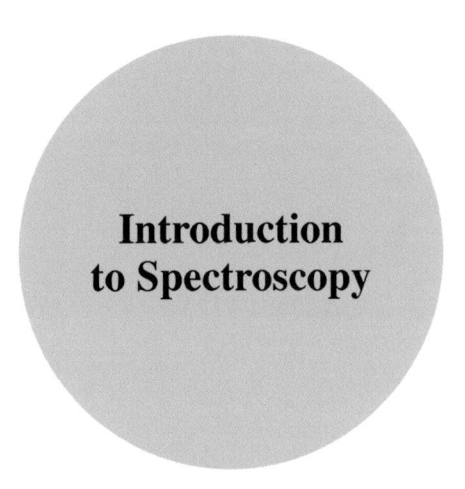

Introduction to Spectroscopy

General Spectroscopy Theory

Introduction

Anyone who has looked at a rainbow has surely found them to be interesting and beautiful. For a long time few people understood what they saw. It was Sir Isaac Newton (1642–1727) who demonstrated what causes rainbows and some of the deeper characteristics of light. Newton took a glass prism and allowed a sliver (light through a small slit) of light to fall on it. What was seen on the other side of the prism was a rainbow of colors. White light was broken down into rainbow colors (Fig. 1.1).

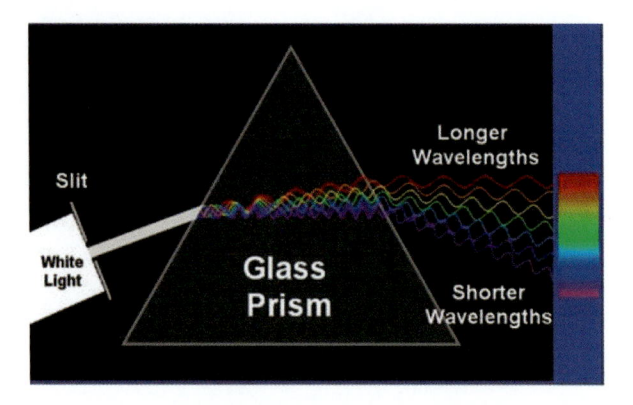

Fig. 1.1 A spectrum produced with a prism

Newton didn't stop there. He then focused the rainbow of colors with a lens and passed the resulting beam through another prism, which then produced the original beam of white light. He carried this further with his famous blue light experiment. He blocked all the rays (colors) coming out of the first prism except the blue portion. He then fed the blue ray through the lens and into the second prism. What came out of the second prism was not white light, but the unchanged blue ray. From this he deduced that light is composed of many single colors and when put together produced white light. This was the beginning of spectroscopy.

Physicists continued to experiment with light. Experiments were performed with light produced by different elements. When heated different elements gave off different colored light. By examining this light it was discovered that each element has a unique spectrum, a unique set of bright lines. For example, holding salt crystals (sodium chloride) in a flame produces a yellow colored light. Detailed examination of the yellow light showed two very bright yellow lines. These are now known as the sodium D lines (with wavelengths of 5889.950 and 5895.924 Å).

Electromagnetic Spectrum

Analysis of spectra is not limited to the visible portion of the electromagnetic spectrum. There are spectral lines of interest that range above the visible bands into the ultraviolet and X-ray regions down the other side through the far infrared band into the radio frequency region (Fig. 1.2).

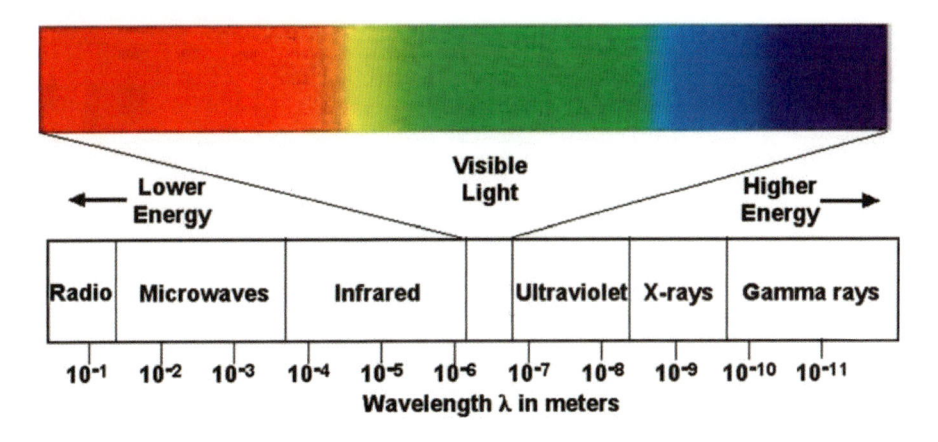

Fig. 1.2 Electromagnetic spectrum

Kirchoff's Laws

In the mid-1800s German physicist Gustav Kirchoff (1824–1857) experimented with the spectra of light. He formulated three empirical rules of spectra analysis:

Rule 1
A hot opaque solid, liquid or gas will emit a continuous spectrum when under high pressure.

Rule 2
A hot gas under low pressure (i.e. much less than atmospheric) will emit a series of bright lines on a dark background. Such a spectrum is called a bright line or emission spectrum.

Rule 3
When light from a source that has a continuous spectrum is shone through a gas at a lower temperature and pressure, the continuous spectrum will be observed to have a series of dark lines superimposed on it. This kind of spectrum is known as a dark line or absorption spectrum.

For many years scientists did not understand how the bright and dark spectra lines were created. Almost all the ideas were from empirical efforts.

Energy States

It wasn't until 1913 when Danish physicist Niels Bohr (1885–1962) managed to explain the spectrum of the element hydrogen. In a simplified model of the hydrogen atom the atom's nucleus is a single proton and is orbited by a single electron. The atom can have different energy states where the electron's orbits are different. When the orbit changes from a higher or more energetic orbit to a lower or less energetic orbit, a photon is emitted with energy equal to the difference of the energy levels. When a photon of the right frequency interacts with a hydrogen atom's electron, the electron's energy state is raised and the photon is absorbed. Only discrete jumps are permitted and none between these specific levels. This means only photons of specific wavelengths or frequencies can be absorbed and emitted (Fig. 1.3).

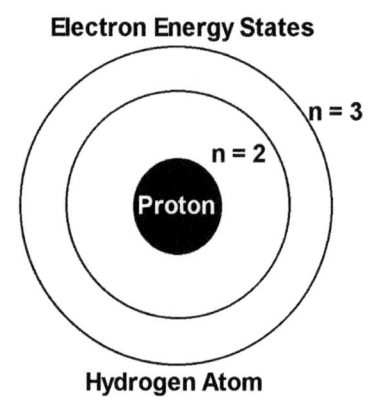

Fig. 1.3 Hydrogen energy states

According to the Bohr model of the hydrogen atom, electrons exist in quantized energy levels. These energy levels are described by a principal quantum number **n**. where **n** is an integer 1, 2, 3, ...

Electrons may only exist in and may transit between these states. The energy (**E**) of the photon emitted or absorbed is equal to Planck's Constant (**h**) times its frequency (**v**).

$$E = h * v$$

Planck's Constant $h = 6.57 \times 10^{-27}$ **erg** $- sec$

Frequency **v** is inversely proportional to wavelength and equal to the speed of light **c** divided by the wavelength λ.

$$v = c / \lambda$$

Thus

$$E = h * c / \lambda$$

Because of the extremely high frequencies of light, photons are usually described by their wavelength in Ångströms (Å) or nanometers (nm). Remember one angstrom is equal to 0.1 nm or 1.0×10^{-10} m. The hydrogen alpha (Hα) line is 6562.81 Å or 656.381 nm.

Atomic Hydrogen

Hydrogen has several series or sets of energy states, the Lyman, Balmer, Paschen, Brackett, Pfund, Humphreys Series (Table 1.1 and Fig. 1.4).

Table 1.1 Hydrogen series

Series	n	λ(nm)	Series	n	λ(nm)
Lyman	2	122	Brackett	5	4,050
	3	103		6	2,630
	4	97.2		7	2,170
	5	94.9		8	1,940
	6	93.7		9	1,820
	∞	91.1		∞	1,460
Balmer (α)	3	656	Pfund	6	7,460
Balmer (β)	4	486		7	4,650
Balmer (γ)	5	434		8	3,740
Balmer (δ)	6	410		9	3,300
Balmer (ε)	7	397		10	3,040
	∞	365		∞	2,280
Paschen	4	1,870	Humphreys	7	12,372
	5	1,280		8	7,503
	6	1,090		10	5,129
	7	1,000		11	4,673
	8	954		13	4,171
	∞	820		∞	3,282

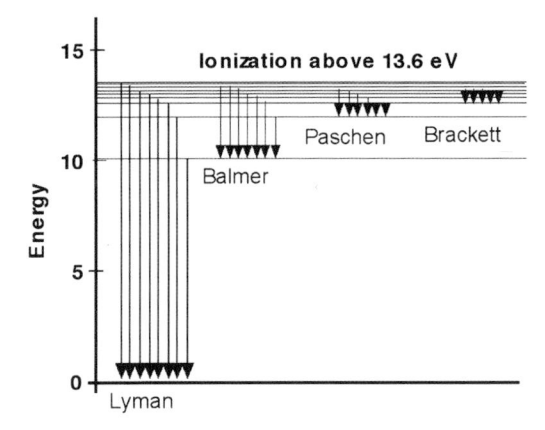

Fig. 1.4 p/o Hydrogen transition series

Balmer Series

In the Balmer Series, the transition from n=3 to n=2 is called the hydrogen alpha (Hα) transition. The Hα line has a specific wavelength of 6,562.81 Å or 656.281 nm. This is visible in the red part of the electromagnetic spectrum as a bright red line (Fig. 1.5).

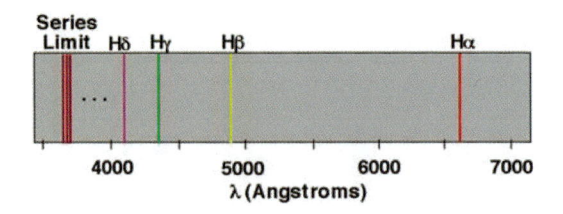

Fig. 1.5 Hydrogen Balmer spectrum

Since it takes nearly as much energy to excite the hydrogen atom's electron from n=1 to n=3 as it does to ionize the hydrogen atom, remove the electron, the probability of the electron being excited to n=3 without being removed from the atom is very small. Instead, after being ionized, the electron and proton recombine to form a new hydrogen atom. In the new atom, the electron may begin in any energy level, and subsequently cascades to the ground state (n=1), emitting photons with each transition. Approximately half the time, this cascade will include the n=3 to n=2 transition and the atom will emit Hα light. This is why the Hα line is so prominent.

While the hydrogen alpha line appears to be a single line it is actually a close doublet with lines at 6,562.84 and 6,562.72 Å. The lines have different intensities so a correct wavelength (6,562.81 Å) that is a weighted average of the intensities is used.

Types of Spectra

There are three basic types of spectra, Continuous, Emission and Absorption Spectra. There are also spectra that contain all three types (Fig. 1.6).

General Spectroscopy Theory

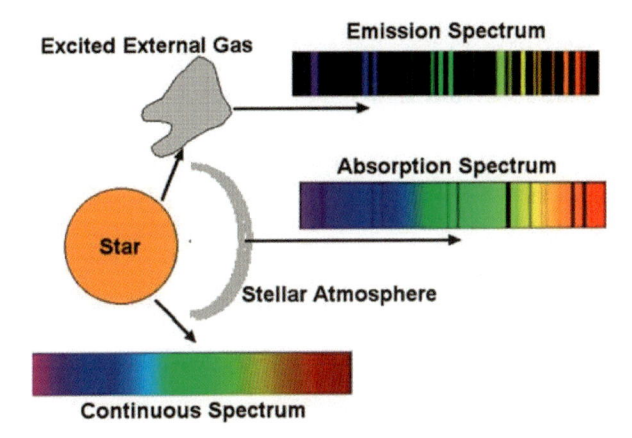

Fig. 1.6 Types of spectra

Continuous Spectra

A Continuous Spectrum such as that produced by the Sun or other stars is the result of very hot gases containing atoms with high kinetic energy in collision. While elements produce their discrete spectral lines, the lines get blurred and as a result a continuous spectrum without individual emission lines being seen (Fig. 1.7).

Fig. 1.7 Continuous spectrum

Emission Spectra

An Emission Spectrum is produced when atoms are less excited than with the Continuous Spectra. An individual atom's unique spectral lines or the lines of many different elements can be seen as the atom's electron states drop to lower levels and emit photons. These lines are seen against a dark background (Fig. 1.8).

Fig. 1.8 Emission spectrum

Absorption Spectra

When photons from a continuous spectrum pass through a gas made up of one or more elements, the specific lines for each element will absorb photons of the wavelengths for those specific elements. This produces a continuous spectrum with holes in it, dark lines where photons of specific frequencies are absorbed. These are the Fraunhofer lines seen in the spectrum of the Sun. The Sun produces a continuous spectrum, but as the light goes through the Sun's atmosphere, elements in the atmosphere absorb photons that are of their specific energy levels, i.e., wavelengths (Fig. 1.9).

Fig. 1.9 Absorption spectrum

Astronomical Spectroscopy Theory

Introduction

Since the spectrum of a light source can tell so much about the source, spectroscopy is ideal for studying light from distant astronomical objects.

All stars are fueled by nuclear fusion where hydrogen atoms are fussed to form helium atoms and give off enormous amounts of energy in the form of radiation. It is this radiation that we see as the starlight. Stellar spectroscopy is the study of the spectra of that starlight. Much can be learned about distant stars by analyzing a star's spectrum. Stars are in different stages of evolution and thus produce very different spectra. Remember, like our Sun, other stars will produce a continuous spectrum and like our Sun the star's atmosphere acts like a filter and allows us to determine what elements are in the atmosphere as well as a host of other information. Some stars produce significant emission lines too.

In most cases multiple lines for a given element can be seen. The lines may be shifted slightly up or down in wavelength relative to a standard stationary laboratory wavelength for that element. This shift is due to the star's motion and other phenomena. When seen, this is sometimes an indication of a multiple star system even though only one star can be resolved visually from Earth. When two sets of lines for a given element are seen this is known as a spectroscopic binary star system. By analyzing the spectral lines much more information about the star system can be learned. The spectroscopic binary systems are also good candidates for photometric study as they may be eclipsing binary star systems.

Astronomical Spectroscopy

Fraunhofer Lines

In the early 1800s German physicist Joseph von Fraunhofer (1787–1826) noted that the expanded spectrum of the Sun had some dark lines in it. These are now known as Fraunhofer lines. Also during the early 1800s British astronomer William Hyde Wollaston (1766–1828) used a prism and observed that the Sun emitted a continuous spectrum that had 784 dark lines. These were the lines Fraunhofer had noted. Fraunhofer realized that some of these dark lines were at the same position in wavelength as bright emission lines of spectra of various elements, which were studied in the laboratory. It is now known that there are thousands of these lines and they represent 67 different elements found in the Sun (Fig. 1.10).

Fig. 1.10 Fraunhofer lines in a solar spectrum

Line Broadening

The shape of a spectral line is influenced by a number of processes in the stellar atmosphere. According to its intensity a line's shape or profile is described as being weak, strong or very strong.

The three main processes that affect the shape of a spectral line are collisional broadening, Doppler broadening and rotational broadening. In addition lesser effects called the Zeeman and Stark effects can also cause splitting of the spectral lines.

Collisional Broadening

If two atoms collide, the electrons of each atom will repel each other and distort their respective energy levels. If a collision happens when one of the electrons is interacting with a photon, then the photon's energy will be altered from the value it would have had for the atom in an undisturbed state. In a gas that is at a moderate temperature and density, collisions between atoms are infrequent and so interaction is likely to happen when the atom is undisturbed. Higher temperatures and pressures cause the photon energies to vary over a considerable range. This spread of energies relates to a spread of frequencies/wavelengths and causes the spectral line to widened or broadened.

Doppler Broadening

Due to their thermal energy, atoms in a star's atmosphere have random velocities. At any instant some of the atoms travel towards us while others are moving away when they emit or absorb photons. This produces a Doppler shift/broadening of the emission or absorption lines. Doppler broadening is a lesser effect than collisional broadening.

The approximate broadening is given by:

$$\Delta\lambda = \lambda_o * v / c$$

Where the change in the wavelength is $\Delta\lambda$, where λ_o is the stationary spectral line wavelength, **v** is the velocity of the atom that interacted with the photon and **c** is the velocity of light.

Rotational Broadening

A star that is rotating will produce a Doppler shift of each absorption line of the star's atmosphere's spectrum. The amount of broadening depends on rotation rate and the angle of inclination of the axis of rotation to the line of sight. This effect can be used to calculate the rotation rate of the star. Assuming the axis of rotation is perpendicular to the line of sight with Earth. If the change in wavelength of a line at wavelength λ is $\Delta\lambda$, then the velocity **v** of atoms on the limb of a rotating star is given by:

$$v = c * \Delta\lambda / \lambda$$

If we know the radius **R** of the star then the period P_r of rotation can be calculated from:

$$P_r = 2 * \pi * R / v$$

Astrophysicists have found that, in general, the hottest stars (type O and B) rotate the fastest with periods as fast as 4 h. G-type stars like the Sun rotate fairly slowly at about once every 27 days.

Zeeman Effect

Electrons in atoms are moving charges that constitute rings of electric current. This produces a magnetic field similar to that of a bar magnet. This Zeeman effect splits the spectral lines. It is possible to relate the degree of splitting to the strength of the star's external magnetic field. From this astrophysicists can obtain information about a star's magnetic field distribution. Zeeman splitting is particularly useful in the study of sunspots, which have very intense magnetic fields and produce pronounced splitting in the absorption spectrum of the sun.

Stark Effect

Like the magnetic field of the Zeeman Effect, an electric field produces a similar spilt and is called the Stark Effect. This is also known as pressure broadening.

Stellar Spectra Classification

The changes in intensity of the hydrogen lines with temperature allow creation of a spectral classification system. The first person to attempt to do this was the Italian astronomer P.A. Secchi who in 1860 classified stars into four distinct groups based on their spectral features. The modern star classification scheme is called the MK system (devised by W.W. Morgan, and P.C. Keenan). A classified star has been assigned a Spectral Type and Luminosity Class.

Wien's Law

By noting the wavelength of the peak intensity of the star's spectrum, the temperature of the star can be determined and thus the classification (Fig. 1.11).

Wein's equation: $(\lambda_{peak}/m) = 2.90 \times 10^{-3}/(T/K)$

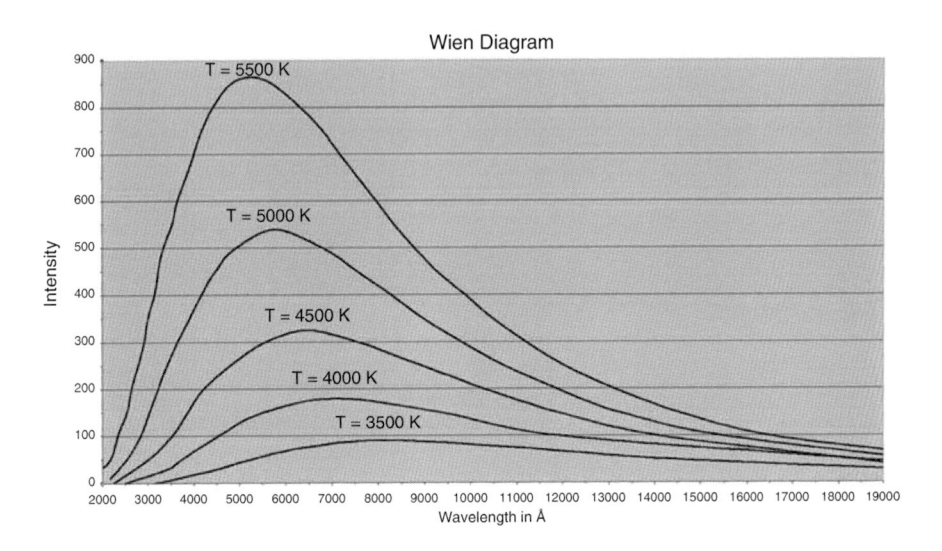

Fig. 1.11 Wien's displacement law

Planck Curves

From Wien's displacement law we know that the black body or Planck curve peak can indicate what the temperature of the body is. The shorter the wavelength where the peak occurs, for the visible spectrum (4,000 Å to 7,000 Å) the range is from 4,500° K to 5,500° K. Each star's spectrum can produce a Planck curve. The peak of that Planck curve relates to the temperature of that star.

Spectral Type

The spectral type of a star is designated by one of seven letters O, B, A, F, G, K, M, starting with the hottest type (O type) to the coolest type (M-type).

The classification of stars into spectral types is actually more complex. Each type can be subdivided into at least 10 subdivisions so a star of type A5 is lying halfway between type A0 and F0. The spectral type order (OBAFGKM) is commonly remembered by the mnemonic 'Oh Be a Fine Girl (or Guy) Kiss Me!' The table below shows the temperatures and characteristic features in the star's spectrum that distinguish spectral types (Table 1.2).

Table 1.2 MK of spectral type

	Surface Temp/K	Spectral type
O	³33,000	Ionized helium (He II)
B	10–33,000	Neutral helium, hydrogen lines start to appear weaker
A	7,500–10,000	Strong neutral hydrogen (Balmer lines) visible
F	6,000–7,500	Ionized calcium (Ca II) visible, hydrogen lines
G	5,200–6,000	Ionized Ca II very prominent, much weaker neutral H lines, also other metallic lines such as Iron (the sun is a G-type star)
K	3,700–5,200	Neutral metals such as Ca and Fe prominent, molecular bands visible
M	£3,500	Molecular bands very visible, particularly those of Titanium Oxide (TiO)

Luminosity Classes

For a given temperature, some stars are more luminous than others. This is usually because the star is larger and its outer atmosphere more tenuous and at a lower pressure than a fainter star. The spectral lines of very luminous stars are much narrower since the effect of line broadening due to collisions is much less and the line profile is sharper.

Astronomical Spectroscopy Theory

Stars can be further classified for each spectral type in terms of luminosity on the basis of the 'sharpness' of their spectral lines. These luminosity classes are denoted by Roman numerals and are divided into seven star-types (Table 1.3):

Table 1.3	Luminosity classes
I	Supergiant Stars
II	Bright Giant Stars
III	Giant Stars
IV	Subgiant Stars
V	Main Sequence Dwarf Stars
VI	Sub Dwarf Stars
VII	White Dwarf Stars

Some luminosity classes, particularly those of the supergiants, are subdivided into suffixes **a**, **ab** and **b** and a class written as III-IV means a star with characteristics midway between the two classes.

The full spectral classification thus consists of:

[spectral type] [number] [luminosity class] [suffix (if any)].

For example, the sun is classified as a G2V star. Betelgeuse, a red giant is classified M2Iab. Epsilon Aurigae is classified F0I (Table 1.2).

Various Different Type Stellar Spectra

The following spectra line profiles were created from the RSpec Spectral Library (Figs. 1.12, 1.13, 1.14, 1.15, 1.16, 1.17, 1.18, 1.19 and 1.20).

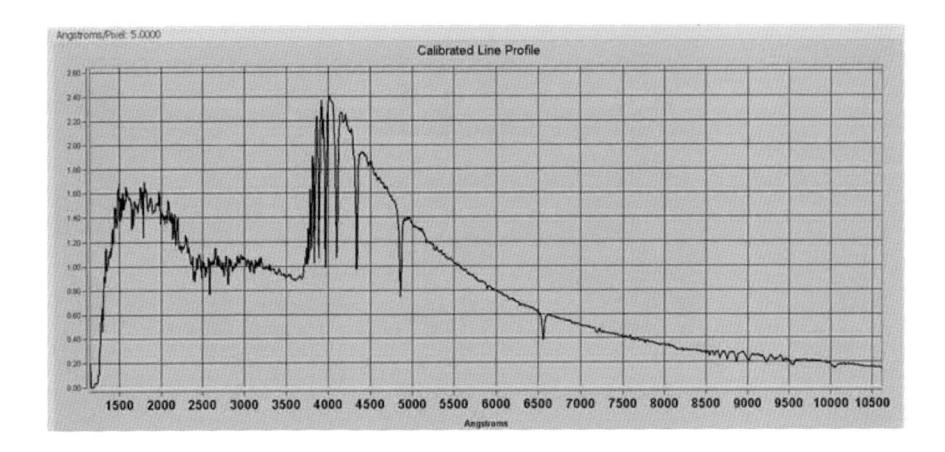

Fig. 1.12 A0V star spectrum

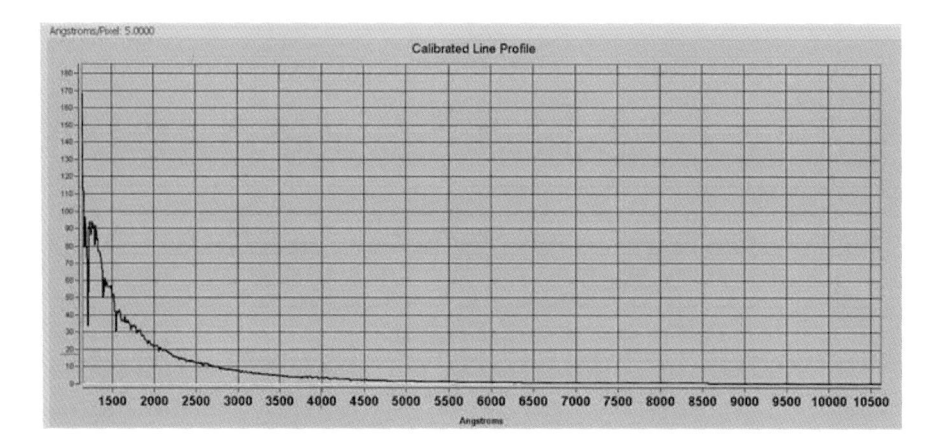

Fig. 1.13 B0V star spectrum

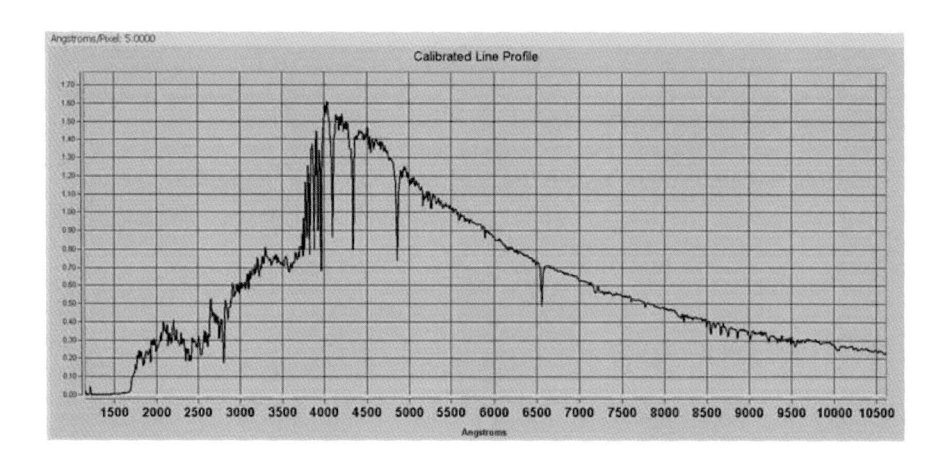

Fig. 1.14 F0V star spectrum

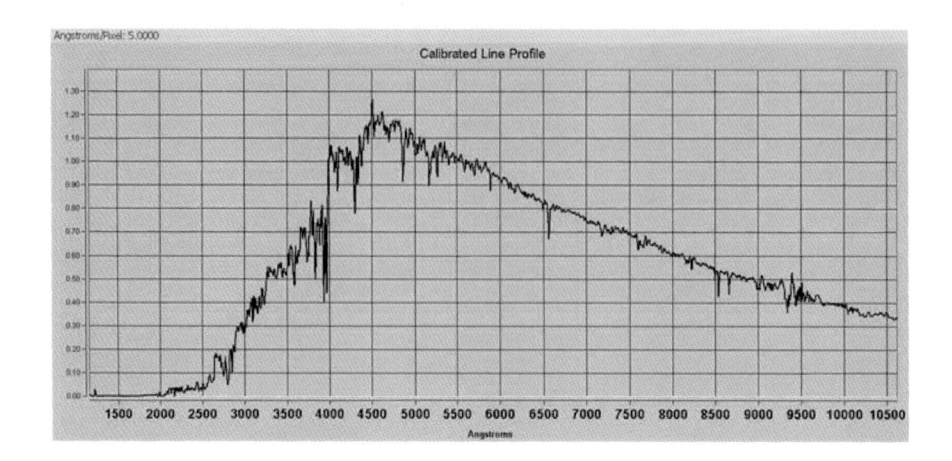

Fig. 1.15 G0V star spectrum

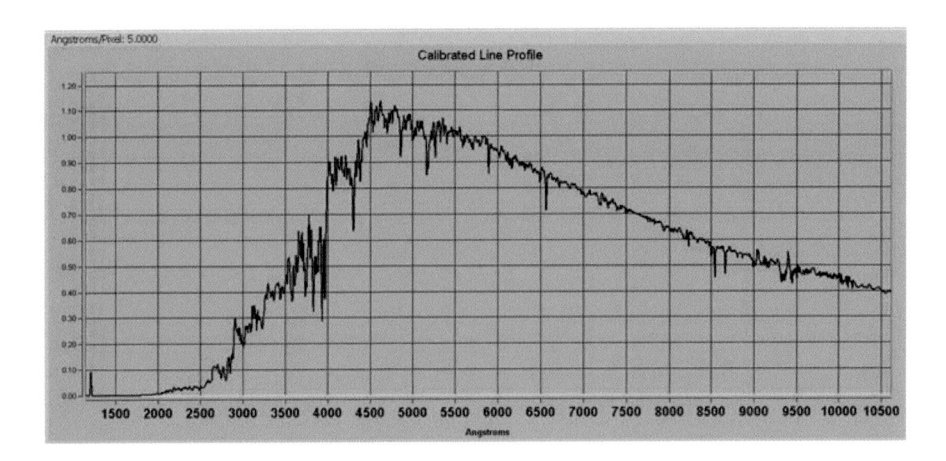

Fig. 1.16 G5V star spectrum

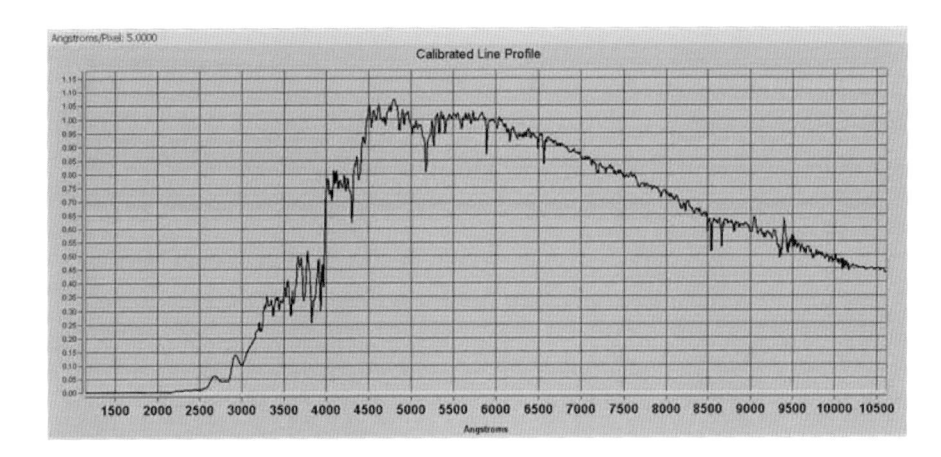

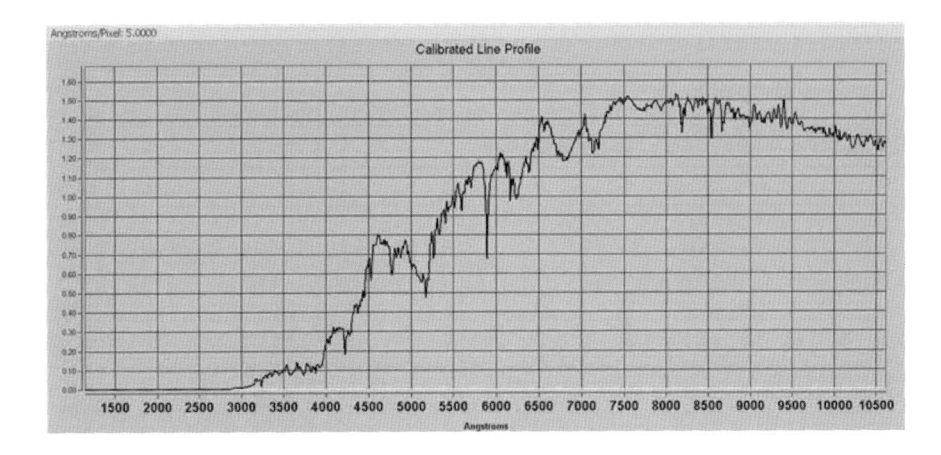

Fig. 1.17 K0V star spectrum

Fig. 1.18 M0V star spectrum

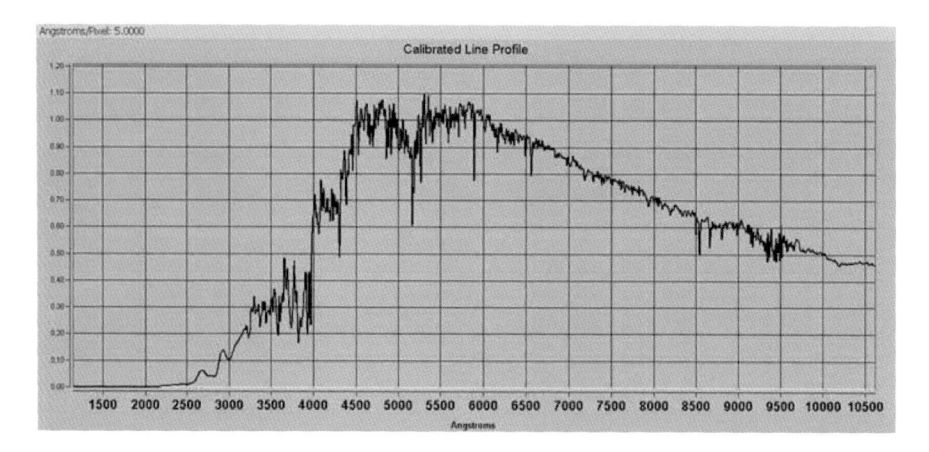

Fig. 1.19 RK0V star spectrum

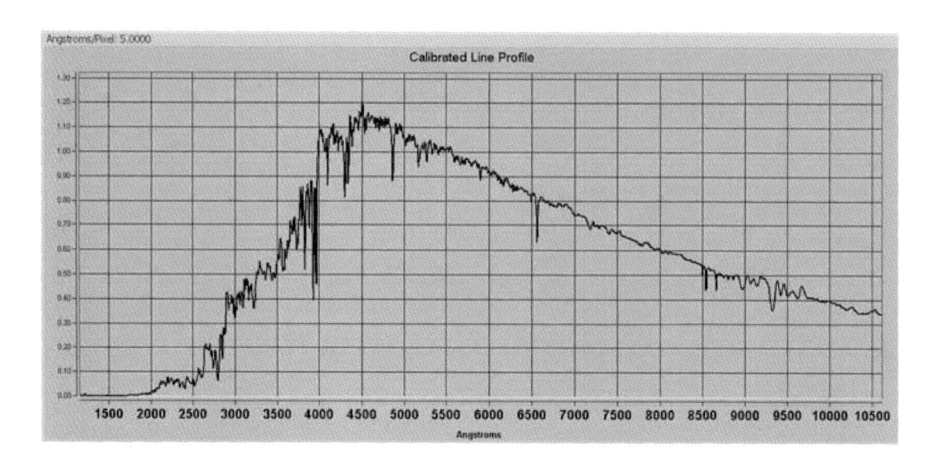

Fig. 1.20 WG0V star spectrum

Chemical Composition

The dark spectral lines observed in a star's spectrum arise from the chemical elements present in the stellar atmosphere. Each element leaves its 'signature' in the form of a pattern of dark spectral lines corresponding to its electron shell structure.

At first it may seem that the more intense the spectral line pattern is, the more of that element the star contains. However, a faint set of absorption lines can be seen because of the high temperature of a star. This means that not all the electrons of an element are in the correct initial energy levels in order to produce a particular line.

In order to calculate the relative abundance of the chemical elements in a star, for a given element, it is necessary to estimate what fraction number of atoms in the first excited state. Then estimate what fraction in the second and so on.

It is found that for the majority of stars, the chemical composition is very nearly the same. By mass, most stars contain about 72 % hydrogen, 25 % helium and the remaining 3 % is made up of other elements (notably iron) in roughly equal abundance.

Astronomical Spectra Processing

Once a star or other astronomical object's spectrum has been imaged, to be of value it must be processed. After doing some image processing the next thing is to process the resulting spectrum. The goal is to produce a wavelength calibrated line profile. More details on creating a line profile from a spectrum image will be provided in Chap. 7, on Software.

Line Profiles

There will be much discussion of line profiles. When astronomical spectroscopy was first being done, the spectrum images were exposed on photographic plates. The plates needed to be carefully developed. Images then had to be studied with a microscope. Now when a spectrum image is taken it is done with a digital camera. The image can still be examined directly, but because the image is now digitized there is a wealth of computer programming available to process the image. No microscope is needed. The processing results in a line profile of the spectrum image. It does not look anything like the spectrum image, but contains a wealth of information. The line profile also makes precise measurements much easier than trying to make the same measurements directly from the spectrum image. All serious science is done from a line profile of the spectrum image.

A line profile is created by first delimiting the spectrum in the image. The image consists of a pixel map with rows and columns of pixels. Each pixel has an intensity number which is an ADU count. The software can then use the delimited area of the image and have all the rows pixel ADU counts for each column in the delimited area summed to produce a total count for a given column. The total ADU count is then plotted on the vertical Y-axis of a graph and the pixel column position plotted on the horizontal X-axis (Fig. 1.21).

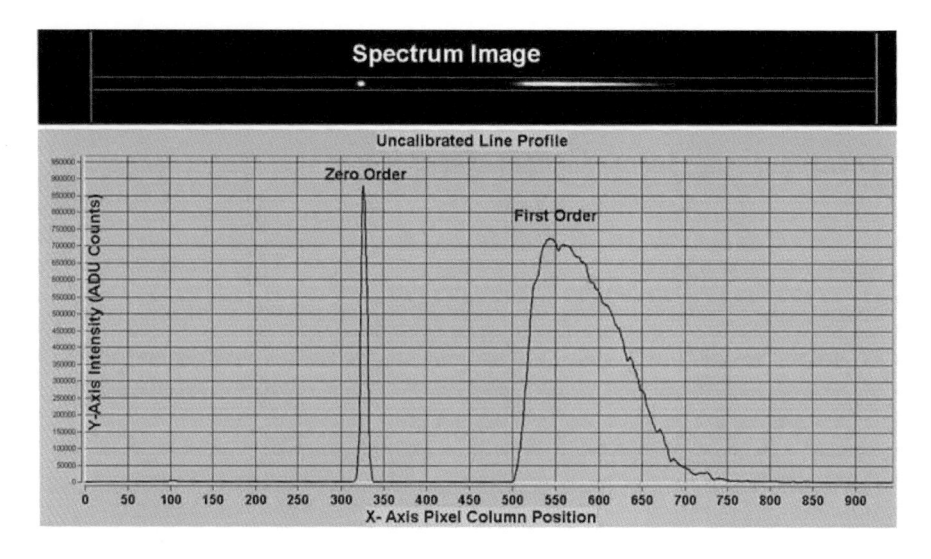

Fig. 1.21 Spectrum image versus line profile

Doppler Effect and Radial Velocities

A spectral line's wavelength can be affected by the relative motion of the star or other object to the observer due to the Doppler Effect. Light from a star will be shifted toward the blue end of the visible spectrum if it is approaching the observer and shifted toward the red end if it is receding. The same equation used for the Doppler Broadening above applies for a Doppler shift due to a radial velocity.

$$\Delta\lambda = \lambda_o * v/c$$

Where **v** is the radial velocity (the velocity of the star as measured along the line of sight of the observer in the manner of a 'radius' drawn from the Earth to the star). A negative **v** (a star moving away from the observer) will produce a shift in wavelength toward a longer wavelength. A positive **v** (a star moving toward the observer) will produce a shift toward a shorter wavelength.

Normalization

To calculate an equivalent width or power of a spectral line, the line profile should first be normalized. While simple in principle this is sometime more difficult than one might expect. To normalize a line profile, the average flux intensity (Analog to Digital Units or ADU counts of the continuum) is divided into the line profile. For a nice smooth continuum this is fairly easy. Many star spectra will produce line profiles that have continuums that are anything but smooth. A good solution

Astronomical Spectroscopy Theory

is to try to determine the edge points of the line of interest and note the continuum value on each side. Average that value and use it for the denominator. This takes a little practice.

The most important thing is if you are doing a study with many of these spectra, to use the same technique for all the data. Once the division is done the continuum is replaced with a line centered at 1.00.

As can be seen in the following image the continuum is not level. Approximate points for Y-axis values at the left and right of the line produce values for Y1 and Y2. The average of these two values can then be used as the continuum value to do the normalization.

In the following case the Y1 is 3920 ADU counts and Y2 is 4059 ADU counts. The average continuum for the line is then 3988 ADU counts. The number 3988 is then divided into the continuum to normalize the line profile. This puts the continuum around the value of 1.0 (Fig. 1.22).

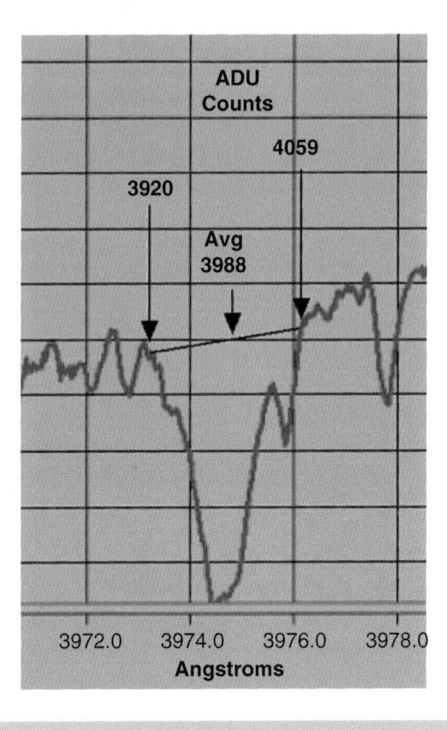

Fig. 1.22 Selected continuum area for normalization

Full Width Half Maximum (FWHM)

Specifying a FWHM for a wave shape allows a description of that wave shape. This is useful when discussing line profiles. To determine the FWHM of a curve the peak is determined, f (max). The width of the curve halfway to the bottom, 0.5 * f (max) is the FWHM (Fig. 1.23).

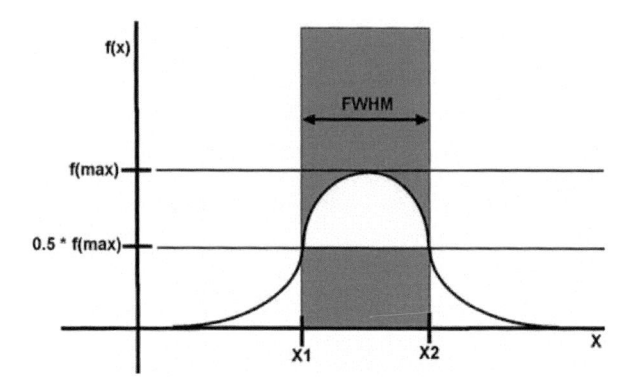

Fig. 1.23 Full width half maximum

Equivalent Width (EW)

Equivalent Width is a measure of the strength of a spectral line. The strength can vary over time and is an important piece of information. Equivalent Width is usually not a concern with low-resolution spectra, but is very important for high-resolution spectra.

Expressing a line profile's Equivalent Widths allows expressing of the line's significance. The area between the curve or profile and the continuum is the EW of that line. The area is equal to the Intensity (continuum normalized to 1.0) times EW in Å.

The following diagram shows an emission spectral curve. There is some debate as to the sign of the EW. In the days of film spectroscopy it seems areas below the continuum (absorption lines) were considered positive while curves above the continuum (emission lines) were considered negative. This is a bit against common sense as usually values above a line are positive and negative below the line. In fact many professional papers in the last few years use the emission EW as positive and absorption EW as negative. The bottom line is it is best to specify what convention is used when reporting EW (Fig. 1.24).

Astronomical Spectroscopy Theory

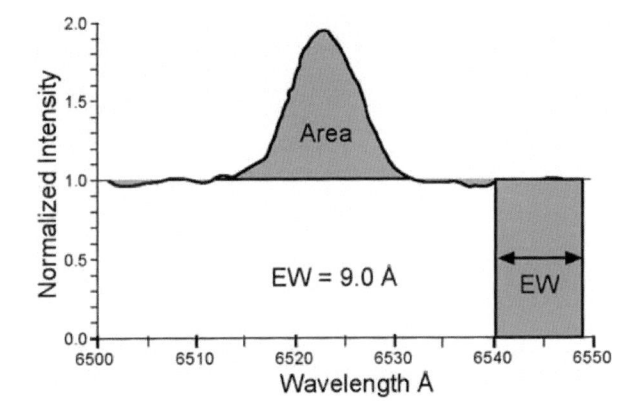

Fig. 1.24 Equivalent width

The following Figure shows a hypothetical hydrogen spectrum. The blue absorption curve has an EW of −0.1 Å; the blue emission curve has an EW of +0.2 Å. The main absorption line has an EW of −0.6 Å and the red emission curve has an EW of +0.3 Å. The values can then be checked against other spectra for changes (Fig. 1.25).

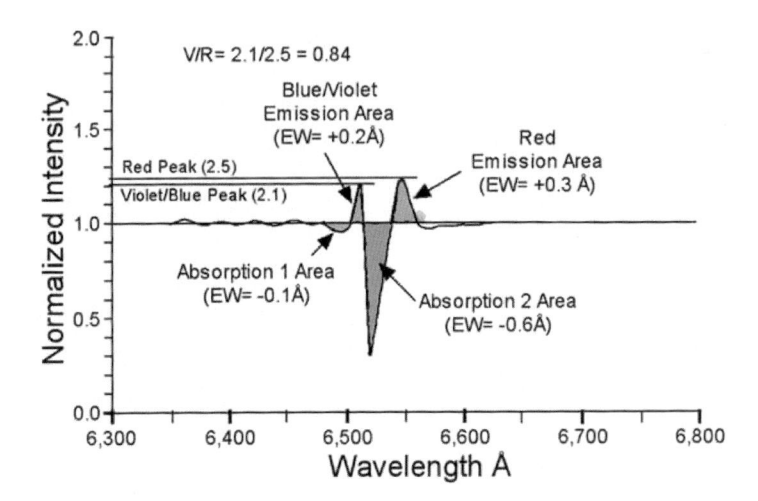

Fig. 1.25 Equivalent widths and VR of hydrogen alpha

V/R Ratio

Along with the EW a term called the VR ratio can be specified. This is the ratio of the violet (V) or blue emission line to the red (R) emission line level. In the above case the V/R is 0.84. These lines may not be actually blue or red, but the color just indicates the wavelength direction of the line, toward the blue or toward the red.

Wavelength Calibration

Once a spectrum has been converted to a line profile it must be wavelength-corrected to be useful. The vertical Y-axis of the line profile is the intensity (ADU count) axis and the horizontal X-axis is the pixel column number. The trick is to associate the X-axis column pixel positions with specific wavelengths. This can be done with a spreadsheet program like Excel, but is much easier done with a specific spectroscopy processing software program, e.g., RSpec or VSpec.

Linear Versus Non-Linear Wavelength Calibration

The spectrum processing software handles all this easily, but some may be interested in knowing more about what is going on.

Linear

In relation to a line profile linear is a straight line. On an X/Y graph the standard equation for a straight line is:

$$Y = mX + b$$

Where **m** is the slope (DY/DX) and **b** is the Y-intercept (where the line intersects the Y-axis when X=0) (Fig. 1.26).

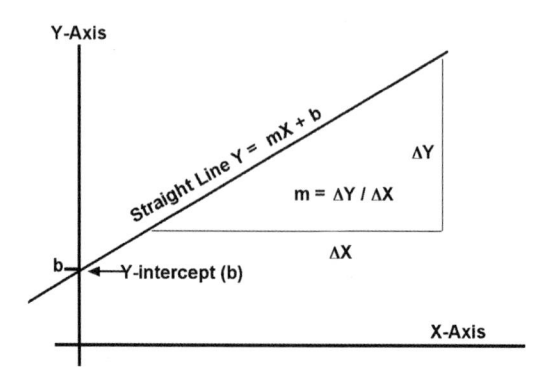

Fig. 1.26 Linear – straight line

This can be a bit confusing as a line profile has an X-axis and Y-axis, but they represent the pixel position/wavelength versus Intensity respectively, whereas the above X-axis and Y-axis represent pixel position and wavelength. If the pixel to wavelength relationship is linear and produces a straight line, knowing just two points can define the line. For example if the following are known:

Pixel Position 1 = 92 and the wavelength is 0 Å

Pixel Position 2 = 572 and the wavelength is 6,500 Å

That means for a pixel position change of 480 (572–92) there is a corresponding wavelength change of 6,500 Å (6500–0). This relates to a dispersion of 13.1 Å/Pixel. Knowing the dispersion and one point, the line profile can then be calibrated. For the above calibration a pixel position of 500 the corresponding wavelength would be 6,563 Å. Note, the first position does not have to be the zero point at 0 Å. It can be some other point on the first order spectrum profile.

The visible spectrum is not linear, however. For low-resolution work a linear calibration can still provide a good approximation. While a multi-point non-linear calibration can be done with a low-resolution line profile, it usually will not significantly change the calibration, even if it is much more accurate.

Non-Linear

The spectrum pixel position vs. wavelength changes in a non-linear way, i.e., the line is not a straight line, but curved. For medium and high-resolution work where more precise wavelengths are needed, this will provide a more accurate calibration. This is more complex than the linear approach, but again, it is handled nicely by software. A non-linear curve is represented by a polynomial in the form of:

$$Y = a_0 + a_1 X + a_2 X^2 + a_3 X^3 + a_4 X^4$$

Again, the X-axis and Y-axis here refer to pixel positions and wavelengths. The order of the polynomial is determined by the highest power of X. In the above example the order is 4. What needs to be done is to calculate the coefficients, a_0, a_1, a_2, a_3 and a_4. This is done easily in Excel, but again this is only shown to explain what is going on and the spectrum processing software handles this with easy. A table of pixel positions and corresponding wavelength is created in Excel (Table 1.4).

Table 1.4 Polynomial pixel versus wavelength table

X-axis pixel position	Y-axis wavelength Å
801	5,320.000
1201	5,890.000
1245	5,970.553
1356	6,070.434
1375	6,090.616
1417	6,140.306
1488	6,210.728
1533	6,260.650
1571	6,300.479
1599	633.4431
1648	6,380.299
1668	6,400.255
1781	6,500.653
1812	6,530.288
1860	6,570.400
1889	6,590.895

To see the graph and display the results a scatter chart must be created (Fig. 1.27).

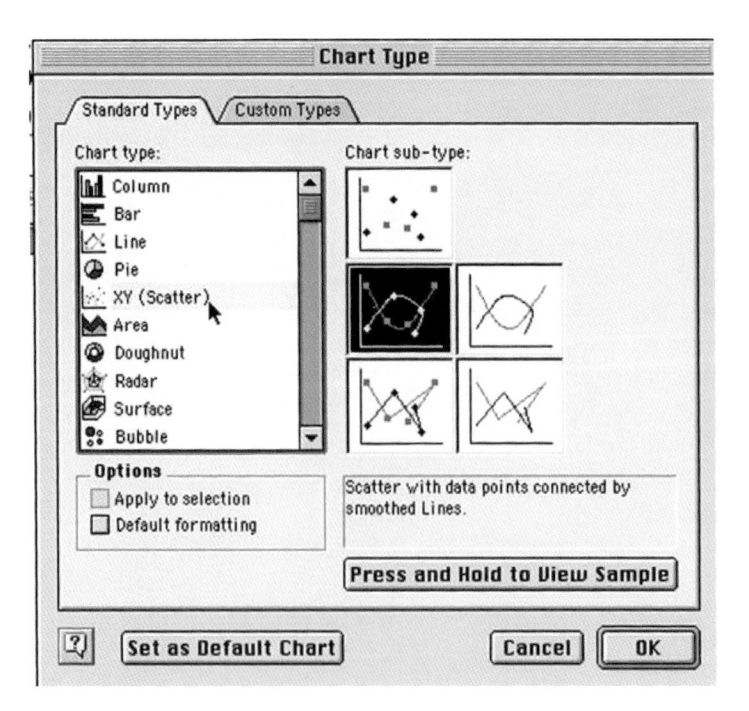

Fig. 1.27 Excel chart type

Astronomical Spectroscopy Theory

Once the chart is displayed, click on the displayed chart and then from the top menu click on "Chart" and the "Add Trendline." Select the Polynomial Trend/Regression Type. If you have 5 or more points select 4 order and then "Okay." Excel allows up to 6 orders (Fig. 1.28).

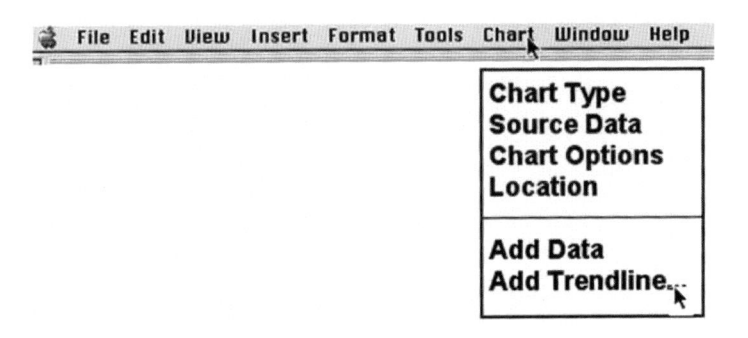

Fig. 1.28 Excel trendline selection

There are several different Trendline selections displayed. Select the Polynomial choice and set the "Order" to 4 (Fig. 1.29).

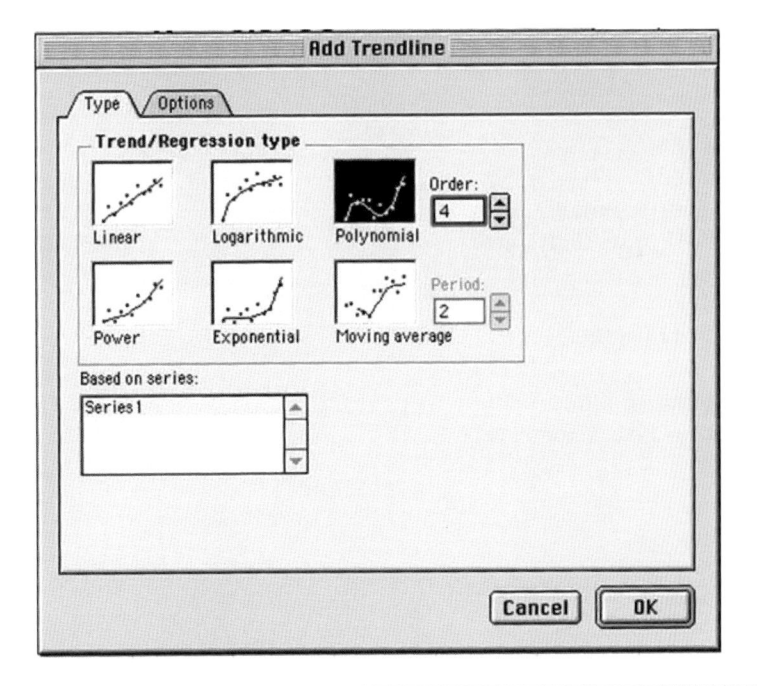

Fig. 1.29 Excel polynomial trendline selection

The polynomial will be displayed with the coefficients entered and a value for R^2 shown. This number is a representation of how good a fit the curve is. A value of $R^2 = 1.0000$ is best (Fig. 1.30).

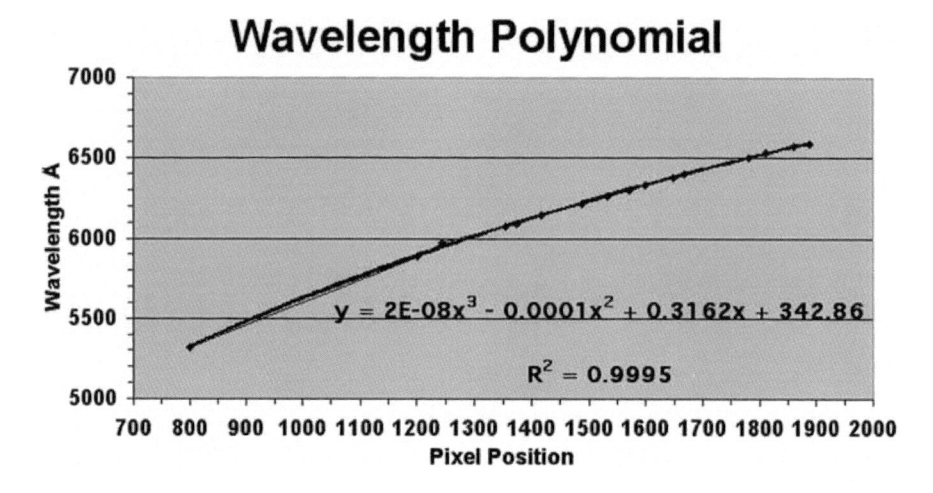

Fig. 1.30 Excel polynomial graph

The coefficients for a 3rd order solution are:

$a_0 = +342.86$
$a_1 = +0.3162$
$a_2 = -0.0001$ (essentially zero)
$a_3 = +eE-08$ (essentially zero)

Un-calibrated Line Profile

When first created a line profile's Y-axis represents Intensity/Flux (ADU counts) and the X-axis Pixel Position (Fig. 1.31).

Astronomical Spectroscopy Theory

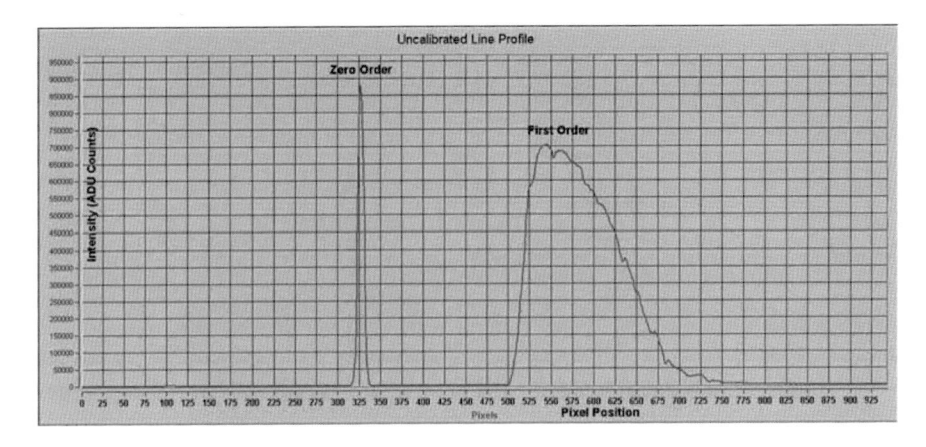

Fig. 1.31 Un-calibrated line profile

Wavelength-Calibrated Line Profile

When wavelength-calibrated the X-axis becomes the wavelength axis in units of Ångströms rather than pixels. The dispersion can now be determined and is displayed at the upper left, for RSpec. The dispersion for this calibrated line profile is 21.4477 Å/pixel. If the optical setup does not change, the dispersion and one other point can be used to provide future calibrations of line profiles (Fig. 1.32).

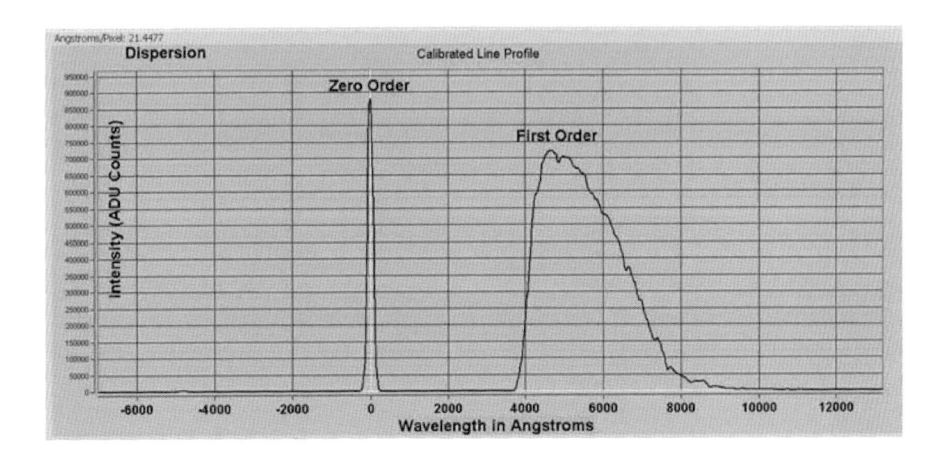

Fig. 1.32 Wavelength calibrated line profile

For low-resolution work where the star is imaged (zero order spectrum) along with the first order spectrum, the zero order point can be used as one of the points. It is at the zero Ångström position. As long as one additional point, such as an identified line in the first order spectrum profile, is determined, the X-axis can be wavelength-calibrated.

One Point Calibration

Once the line profile is wavelength-calibrated the scale or dispersion (Å/pixel) can be determined. If the optical system does not change, that dispersion is still valid for additional image line profiles and only one reference point is needed to do the calibration (Fig. 1.33).

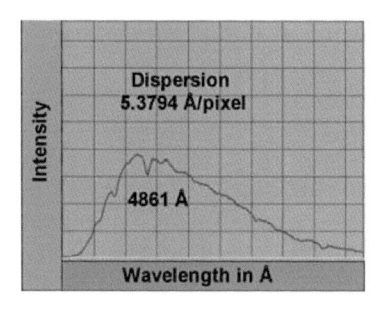

Fig. 1.33 Single point and dispersion wavelength calibration

Multiple Line Calibration

For mid and high-resolution work the zero order spectrum will not be available. Calibrating the narrow window of a high-resolution line profile can be a challenge. Typically neon lines are used as a reference and a two-point or multi-point calibration done. With multiple points higher order non-linear curve fitting polynomials can be used to increate the accuracy of the calibration. At least one more point is needed for the order, i.e., a 4th order polynomial requires at least 5 points. Sometimes telluric lines can be used for additional points (Fig. 1.34).

Astronomical Spectroscopy Theory

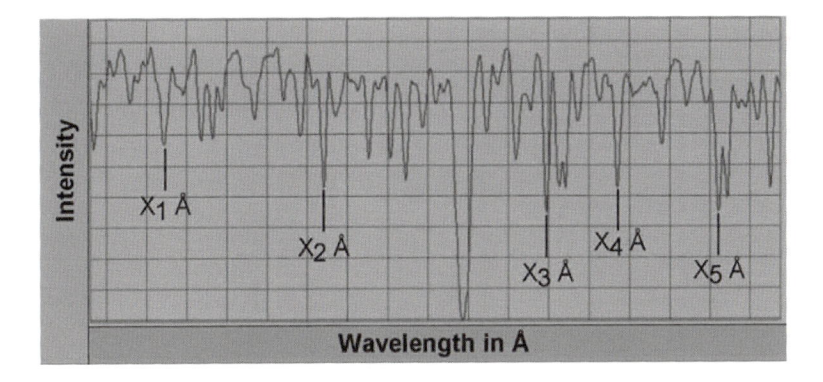

Fig. 1.34 Multiple line/non-linear wavelength calibration

Spectral Order

When a spectrum is created there will be multiple spectra on each side of the zero order spectrum. The brightest of the higher order spectra is the first order spectrum. The higher order spectra become increasingly fainter the farther they are from the zero order spectrum. The zero order spectrum, while called a spectrum is not a spectrum, but the actual image of the source of light. For stellar spectra, the zero order spectrum is the star's image and is located at zero Ångströms. It should be noted that echelle grating type spectrographs are designed to make use of the higher order spectra. The spectral image below shows the right first order spectrum brighter than the left first order spectrum. This is because the grating used has been blazed to make one first order spectrum brighter than the other (Fig. 1.35).

Fig. 1.35 Zero and two first order spectra

Resolution Versus Dispersion

This book will be discussing high, mid and low-resolution spectroscopy. It is important to understand what resolution is.

Resolution and dispersion are sometimes confused. Sometimes resolution is specified as so many Å. While that does indicate a resolution, it does not tell the real story. The resolution Å number is only valid at one region. A resolution of 0.9 Å may be valid at 6,500 Å, but will be different at 4,500 Å. The theoretical spectral resolution **R** is a dimensionless number and defined as:

$$R = \lambda / \Delta\lambda$$

where

λ is the wavelength of a spectral line of interest and $\Delta\lambda$ is the minimum distance between two features.

High-resolution is usually defined as R > 10,000. A Lhires III with a 2,400 l/mm grating has a resolving power of around 17,000 or being able to resolve about 0.38 Å at 6,500 Å. The Star Analyser with its 100 l/mm grating can have a resolution approaching 200.

Dispersion is just the scale factor for the CCD chip. It is expressed as Å/pixel or Å/mm. The Star Analyser has a dispersion of between 5 and 25 Å/pixel for equipment with a ~8.5 mm pixel CCD. Other equipment can have a dispersion of up to 50 Å/pixel. The Lhires III can have dispersion of 0.1 Å/pixel or less. The dispersion relies on the optical setup (mainly distance from the grating to the detector) and camera pixel size. Usually the lower the dispersion the better.

Heliocentric Wavelength Correction

When doing high-resolution spectroscopy and determining radial velocities, a correction for the motion of the Earth is needed. This correction is wavelength dependent and is a number in Ångströms to be added to or subtracted from the wavelength of the line of interest. Depending where the star is and where the Earth is in its orbit around the Sun, the radial velocity of the Earth toward or away from the star can vary between +67,000 MPH going toward the star to −67,000 MPH going away from the star (Fig. 1.36).

Astronomical Spectroscopy Theory

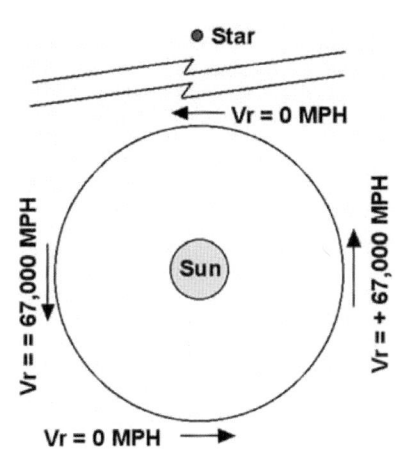

Fig. 1.36 Earth's radial velocity

In addition, depending on the latitude of the observer the Earth's rotation and tangential velocity can contribute up to 1,024 MPH to the radial velocity (Fig. 1.37).

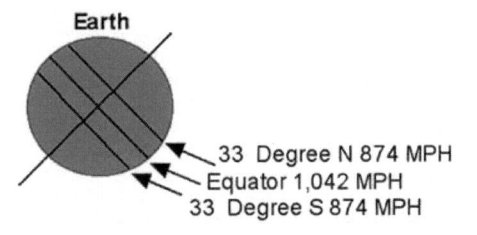

Fig. 1.37 Tangential velocity – 1,042 * latitude

To calculate the heliocentric correction velocity and therefore the heliocentric wavelength correction there are a number of items that must be taken into consideration that determine where the star is relative to the Earth. These are:

Observer's Latitude (e.g., 33.5017)
Observer's Longitude (e.g., 112.2220)
Star's RA (e.g., 18h 36m 56.2s)
Star's Dec (e.g., 38d 47m 01s)
UT Date of Observation (e.g., 06m, 24d, 2008y)
UT Time of Observation (e.g., 07h 32m 15s)

The program used for the calculation determines the exact formats for data entry. Once the profile is wavelength-calibrated, a heliocentric correction can be made. VSpec software, which will be discussed in section "VSpec French Freeware", has the ability to do this calculation.

Absolute Calibration

The Y-axis of a line profile is usually in units of ADU counts. These counts represent the intensity of the light at various wavelengths. It is sometimes of interest to convert this intensity to an absolute calibration or photon flux calibration. The precise techniques are quite involved and beyond the scope of this book, but a brief description of what is involved is presented.

To determine an absolute or flux calibration of a star of interest a standard star is used. Originally Vega and Sirius were used for this. It is important that the extinction be accounted for. If the star of interest and standard star are at the same air mass then the extinction should be close to the same and can be ignored. Otherwise the extinction must be taken into account.

Photon flux is in the units of photons/cm²/sec/nm. The precise equation is:

$$\mathbf{F} = 5.5 \times 10^6 \, (1/\lambda) 10^{-0.4M}$$

Where **F** is the flux (photons/cm²/sec/nm) and **M** is the magnitude at wavelength λ (in nm).

Further investigation is encouraged. The following are a few papers that may help:

Hayes, 1970: 12 standard stars (Mags 0–5)
http://labs.adsabs.harvard.edu/ui/abs/1970ApJ...159..165H
Gunn & Stryker, 1983 Stellar spectrophotometric atlas, 175 stars mags 3–10
http://labs.adsabs.harvard.edu/ui/abs/1983ApJS...52..121G
Bessell et al. (1999), Bright A stars (Mags 3–5)
"Spectrophotometry: Revised Standards and Techniques"
http://labs.adsabs.harvard.edu/ui/abs/1999PASP..111.1426B
Massey et al. 1988, 25 Spectrophotometric Standards (Mags 7–12)
http://labs.adsabs.harvard.edu/ui/abs/1988ApJ...328..315M

Supernovae Classification

Many new supernovae can be easily imaged with a mid and low-resolution spectrograph such as the ALPY 600 and Star Analyser. Significant silicon lines (Si II) seen at around 4,000, 5,200, 5,400 and 6,150 Å will indicate a Type Ia thermonuclear runaway star as opposed to a giant star's core collapse Type II supernova. The Si II 6,150 Å doublet is an excellent indicator.

Modern Astronomical Spectroscopy History

Introduction

To put things into proper perspective a bit of contemporary history of spectroscopy may help. In the late 1880s and early 1900s there was a great thirst for knowledge about the Universe. It was commonly thought that everything that could be seen even with the largest telescopes was within our Universe called the Milky Way. It was obvious the planets of our solar system were close, but how far away were the stars and other astronomical objects? Determining accurate or even ballpark estimates of distances to astronomical objects is very important. At the time the spectrographs in use used plate film for imaging. These were relatively insensitive, even when cooled with liquid nitrogen. Even the brightest stars produced poor spectra using the largest telescopes available. This was an incentive for larger telescopes.

Determining Astronomical Distances

Parallax

One method of determining unreachable distance is by using trigonometry. Anyone who has taken high school trigonometry will remember that finding the distance to an object is a simple trigonometry procedure. This method is known as "parallax." A baseline of known length is constructed and the angle to the object at the ends measured when the object is midway between the end points (Fig. 1.38).

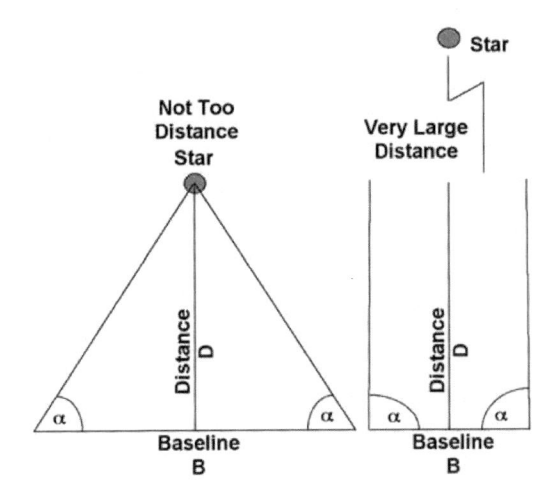

Fig. 1.38 Parallax distance measurements

From basic trigonometry:

$$\text{Tan } \alpha = D/(0.5 \text{ B})$$
$$D = 0.5 \text{ B Tan } \alpha$$

The parallax method works fine up to a point. As long as the Baseline B is big and the angle α is less than 90° the distance can be calculated accurately. For astronomical distances the best that can be done for B is to measure the angle to a star or other object at opposite sides of the Earth's orbit. Even then angle α starts approaching very closely to 90°. This method works for distances up to a few thousand light years, but beyond that angle α becomes too close to 90° to be measured accurately. As the angle α approaches 90° Tan α approaches infinity.

By the late 1800s many interesting astronomical objects other than stars had been discovered. There were globular clusters, big balls of stars, and nebulae, large clouds of apparent gas. While many of these nebulae were indeed cloud like and without much structure, there was another category of nebulae that seemed to have structure. Most astronomers considered these structured nebulae to be part of the Milky Way Universe. There were a few astronomers, however, who were not convinced and felt that these structured nebulae were distant island galaxies like out Milky Way. This was a most important issue that needed to be decided. The key was in determining the distances to these structured nebulae. If these objects were within or close to 100,000, they were part of the Milky Way. Since these objects were too far for parallax measurements another method for determine distance had to be used.

Inverse-Square Law

There are two magnitudes associated with stars, the apparent and absolute visual magnitudes. The apparent magnitude is the visual magnitude observed at the Earth, as it would be measured outside the Earth's atmosphere. This would be the star's corrected V band photometric magnitude. The absolute magnitude is the visual magnitude of a star as would be seen from a standard distance of 10 parsec or 32.6 light years. If we can determine the absolute magnitude of a star and measure it's apparent magnitude, we can determine its distance to the star. The trick is to determine the absolute magnitude of a star. If the star type is known, it is possible to assume an approximate visual magnitude for it. To know the star type requires knowing its spectrum. Until larger telescopes were available, it was not possible to get spectra of the fainter stars. The giant red star Betelgeuse in the upper left of Orion has an absolute visual magnitude of −5.6 (as seen from the standard distance) making it a very bright star. As viewed from Earth some 520 light years away it is still a very bright star, but now with a magnitude of only 0.4 (Fig. 1.39).

Modern Astronomical Spectroscopy History

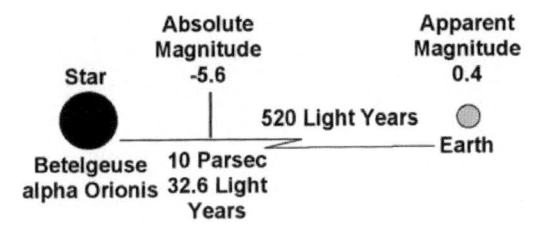

Fig. 1.39 Absolute and apparent magnitudes

How does knowing the absolute and apparent magnitude help in determine the distance to the star? Some basic geometry helps. A radiating spherical source will produce an intensity of 1 unit at the surface for a unit area that is decreased by an inverse-square function.

The area of a sphere is:

$$\mathbf{Area = 4\pi r^2}$$

Where r is the radius of the sphere.

A unit of area at the surface of a sphere can represent the total intensity in that area is (Fig. 1.40):

$$\mathbf{Unit\ Surface\ Intensity = S/(4\pi r^2) = 1\ unit}$$

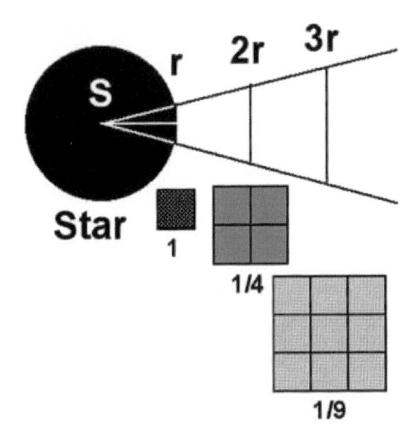

Fig. 1.40 Inverse-square law

This means for every increase in unit distance, Intensity is decrease by the square of the distance. For a distance of 2 radii the intensity is 1/4. At a distance of 3 radii the intensity is 1/9 and so on. Thus if we know the initial intensity or absolute magnitude and the observed intensity or apparent magnitude we can determine the distance. Again, the trick is figuring out the star's absolute magnitude.

Period-Luminosity Relationship

The study of variable stars was the key to determining the absolute magnitudes of some stars. There are basically two types of variable stars, extrinsic and intrinsic variables. The variability of extrinsic variables is usually due to the star being part of a binary star system where the orbits of the stars are on a plane that can be seen from Earth. While only one apparent star can be seen, as one star passes in front of the other the brightness of the star system decreases. If the stars are different colors and/or sizes the eclipse depth will be different in different bands and between the primary and secondary eclipse. Intrinsic variable stars vary because of some phenomena in or on the star's surface. RS CVn stars are famous for their star spots. As the spots, like giant sunspots, rotate around on the star's surface the star's brightness varies. Other stars vary in brightness due to a pulsation of the star. As the star increases in size, it appears brighter. When is falls back, it appears dimmer. One class of these pulsating stars is known as Cepheids. There is a distinct relationship between the period of these stars and their absolute magnitude (Fig. 1.41).

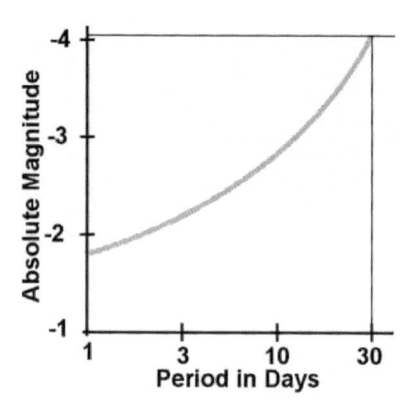

Fig. 1.41 Cepheid magnitude-period relationship

Many measurements were made of close Cepheid variable stars. A relationship was determined between the periods of variation and the star's absolute magnitude. Since the period would not be affected by distance, finding Cepheids in the structured nebulae would provide a means of determining the star's absolute magnitude and thus distance. The problem was again that the telescopes available in the late 1800s were not of sufficient size to resolve stars in these nebulae.

Telescopes

Introduction

To find some Cepheids in the structured nebulae and solve some of the mysteries of the Universe it was obvious that astronomers needed bigger and better telescopes. Perhaps no one knew this better than astronomer George Ellery Hale. Hale was born in Chicago 29 June 1868 and was very active in his early adult life with solar astronomy. He became interested in the Universe size debate and realized the answer lay with new and bigger telescopes. His first endeavor resulted in the Yerkes 40″ refractor.

The Yerkes Observatory

Hale was responsible for the construction and use of the world's largest refracting telescope, the 40″ Yerkes refractor telescope. While there was pressure to build an observatory for the telescope in Chicago, preferable on the University of Chicago campus, it was obvious that would be a very poor site. Instead, in 1897 the Yerkes Observatory was constructed about 80 miles Northwest of Chicago on Geneva Lake in Williams Bay, Wisconsin. Charles T. Yerkes for whom the observatory and telescope is named was the financer. While the telescope was productive it became obvious that an even larger telescope was needed. It was also obvious that the location of the Yerkes Observatory, even in a darker location than Chicago, was far from ideal due to the weather (Fig. 1.42).

Fig. 1.42 40″ Yerkes refractor telescope

The Mount Wilson Observatory

The 40″ Yerkes refractor is gigantic, but at the size limit for a refracting lens. To get around this Hale's next project was a 100″ reflecting telescope. Many people said a 100″ mirror cold not be made and indeed it took many years to compete a successful mirror. When complete, the telescope located 5,500 ft above sea level atop Mount Wilson just above Pasadena, California. With the financial help of John D. Hooker, the Mount Wilson Observatory 100″ Hooker telescope was put into operation in 1917.

It was at the Mount Wilson Observatory's 100″ Hooker telescope that Edwin Hubble became the first person to determine the distance to the Andromeda Galaxy using Cepheid variables stars that he found in Andromeda. The distance turned out to be several times greater than the size of the Milky Way thus proving Andromeda was not part of the Milky Way, but a separate island far away. This had profound effects on astronomy. With the world's largest telescope Hubble went on to use spectroscopy to determine the radial velocities of these galaxies. He determined that the further away a galaxy was, the faster it was moving away. This became what is known as the famous "Hubble's Law." This was very disturbing to Albert Einstein who said because of that his Cosmological Constant was wrong, and it was the biggest blunder of his life. The implication was the Universe was not steady state, but was in fact expanding at an accelerated rate. For the spectroscopy Hubble still used a cumbersome prism-based spectrograph. Photographic plates were used to image the spectra. Hubble pushed the equipment to the limit. What was needed was a still larger telescope (Fig. 1.43).

Fig. 1.43 Author with Mount Wilson 100″ Hooker telescope

The Mount Palomar Observatory

George Hale came to the rescue once again when he pushed the development of the 200″ telescope on Mount Palomar, located 90 miles south of Mount Wilson and just North of San Diego, California. This was a gigantic project with many people saying it could not be done. Hale had proved these people wrong with the 100″ telescope and would do so again with the 200″ telescope. The project was started in 1936 and saw first light on 28 January 1948, nearly 10 years after George Hale passed away. The completed telescope on Mount Palomar was named the Hale Telescope. The Hale Telescope reigned as the largest telescope in the world until 1976. Edwin Hubble continued his observing using the new Hale Telescope. For his efforts and accomplishments the first space telescope was named after him, the Hubble Space Telescope (Fig. 1.44).

Fig. 1.44 Mount Palomar 200″ Hale telescope

The First Astronomical CCD Cameras

The Mount Palomar Observatory and Hale Telescope have many other interesting historical events, one being the introduction of digital cameras for astronomy.

In the late 1960' interest in computers was on the increase. One area that was under investigation was digital memories. In 1969 at the AT&T Bell Labs, William Boyle and George E. Smith invented what was first called the Charge-Bubble Device. This was designed as a digital memory device. Later it became known as a Charge-Coupled Device or CCD. In the 1970s the CCD was recognized as having potential for video recording. The first commercial digital video devices were introduced in 1974.

The Cold War was raging hot during the 1960s and 1970s. The United States kept track of its foes with high-flying reconnaissance jets. This worked well until missiles advanced to the point where a jet was shot down from the high altitude. In the mid-1970s the United States developed a super secret reconnaissance spy satellite known as the KH-11 Kennan reconnaissance satellite, code named 101 and Key Note. It many respects it was very similar to the Hubble Space Telescope, but was designed to look toward the Earth rather than at the stars. It used a 2.4-m mirror and had a theoretical resolution of close to 6 in. on the ground from space. This satellite was unique as it used a single chip digital CCD camera with 800×800 pixels instead of photographic film. Previous spy satellites used film that had to be periodically ejected and caught on the way back to Earth. The KH-11 could radio the digital data back to Earth over a secure channel and the data would be available immediately for analysis. One of the ramifications of this was the idea of using a digital CCD camera for astronomical work. The big problem was the difficulty in making the chips and the high cost. Still in 1977 there was an effort to develop a similar device to be used in the Hubble Space Telescope's Wide Field/Planetary Camera (also known as Wiffpic or WF/PC). Texas Instruments was contracted to manufacture the chips. Because of the high defects 25,000 chips were fabricated. Of these only 125 worked. Of the 125 working CCD chips 8 of the best were chosen. To be chosen the CCD chips had to pass strict space environmental tests as well as other challenges. The estimated value for each of these 8 CCD chips was $50,000 or a total of $200,000, a very large sum at the time. Each CCD chip was a 1.2 cm square with a thickness of 1/15 of a sheet of paper.

In 1983 astronomer James Edward Gunn, while working at the Mount Palomar Observatory, developed an astronomical digital CCD camera using rejected chips from the lot of 125. These were working chips that just did not muster the quality for space, but worked fine in an Earthly environment. Gunn fabricated 4 chips into one camera for a total pixel count of 2.56 million pixels. Each pixel had a resolution of 0.33 arc seconds. Gunn named his camera the 4-Shooter. When cooled with liquid nitrogen, the camera became very sensitive and low-noise. It was not long until Gunn mounted the 4-Shooter at the prime focus of the 200″ Hale telescope and pointed it at the zenith. The camera increased the sensitivity of the 100″ telescope by a factor of 100. With tracking off he watched as thousands of stars passed by on the computer screen, as expected, but also thousands of previously unseen galaxies. As another amazing test he pointed the telescope with the 4-Shooter at a blank area between the stars and took a time exposure. Hundreds of new galaxies were seen. The pretty pictures were amazing and inspiring, but that was just the start. It was not long before the sensitive camera would team up with a spectrograph.

Dutch astronomer Maarten Schmidt was born on 28 December 1929. Schmidt had been studying a special class of radio galaxies. These were star like objects that were very bright. On 27 December 1962 Schmidt used a prism spectrograph at the prime focus of the 200" Hale Telescope. This device was called the Hale Prime Focus Spectrograph. The target was a radio galaxy known as 3C 273. This is a 13-magnitude (visual) star like object in the constellation Virgo. The spectrograph used a small photographic place for the spectrum image. After 4.5 h of exposing the spectrum, the plate was developed. The object was visually very bright in the 200" telescope, but the photographic plate was devoid of a spectrum. The mystery began. For such a bright object, a spectrum should have been seen. A little over a month later on 05 February 1963 Schmidt has success in getting a spectrum of 3C 273. However, the spectrum made no sense. It was as if the object had different elements that produced a different spectrum from those known on Earth. Using a bit of detective work Schmidt looked for the hydrogen Balmer lines. Usually the hydrogen alpha line, at 6,563 Å, is very prominent. There was no prominent line anywhere close to the 6,563 Å region. What Schmidt noticed was there were some significant lines that had the spacing of the shorter wavelength hydrogen Balmer lines, but were shifted way out of the blue region and well into the longer wavelength green and yellow regions of the spectrum. Further study showed the hydrogen alpha line, but shifted far into the red area of the spectrum at around 7,600 Å. This relates to a red shift of 0.158 and thus a distance of 2 billion light years. While an amazing distance, this only deepened the mystery. For such a bright and small object, the size of our solar system or smaller, the energy output was beyond imagination, the energy output of a whole galaxy with a 100 billion stars. These enigmatic objects were first called "quasi-stellar radio sources" or QSR. Some of these objects did not emit significant radio radiation so the classification became more general and the objects became known as "Quasars."

It was not longer after James Gunn tested his 4-Shooter CCD camera on the 200" Hale Telescope in 1983 that Maarten Schmidt approached him with the idea of using the 4-Shooter camera with a new diffraction grating spectrograph, called the Prime Focus Spectrograph, to study Quasars. Modern CCD spectroscopy was born.

For those interested in a more detailed and personal history of the above, Richard Preston's book "First Light" is highly recommended.

There were a couple of milestones that helped pave the way for amateur astronomers to do some serious astronomical work. One of the most obvious is the advance in personal computers. Two other milestones are the CCD camera and the thermoelectric coolers. James Gunn's 4-Shooter was probably the most important step. Gunn used liquid nitrogen to cool his CCD chips. This was not only dangerous, but also expensive and complex, well out of reach for the amateur astronomer. In 1834

French physicist Jean Charles Athanase Peltier discovered the thermoelectric Peltier Effect. Basically when current flows between two conductors heat is removed from one and deposited in the other. This causes the first conductor to cool and the second conductor to get hot. This acts as a heat pump. While this was known for over a century, it wasn't until the late 1900s that the full potential of this type device was realized. Most astronomical CCD cameras are now cooled with thermoelectric coolers, sometimes two stage coolers. These devices can cool the CCD by 40° or more. Modern CCD cameras do not need liquid nitrogen or even as some earlier versions used, water-cooling. The secret to doing spectroscopy is having sufficient sensitivity and low noise to image a spectrum. This is accomplished by using larger telescopes and/or sensitive detectors such as the cooled CCD cameras. The cooled CDD camera and powerful inexpensive personal computers are what really broke open the field of spectroscopy for amateur astronomers.

Chapter 2

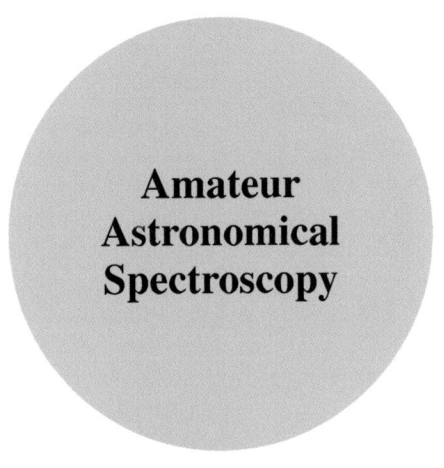

Amateur Astronomical Spectroscopy

Introduction

The goal of astronomical spectroscopy is to produce a wavelength calibrated line profile of the spectrum of an astronomical object, usually a star. From this line profile of the spectrum image, wavelengths can be measured as well as equivalent widths. Stellar and supernova classifications can be done. Doppler shifts can be measured and radical velocities calculated. Elemental lines can also be identified and their strengths measured. All these data can be gleaned from a single image of a spectrum. No wonder spectroscopy is so powerful.

This chapter will discuss the equipment used and techniques for producing a good astronomical spectrum image. Chapter 7 will discuss the software and processing of the spectral images in detail. While certain combinations of equipment will produce better results than others, when starting out with low-resolution spectroscopy almost any combination can be used to produce good first results. Low-resolution work will not provide accurate wavelengths no matter how good the technique. It can be used to identify major elemental lines and even classify supernovae. To determine accurate wavelengths, high-resolution must be used. There is still a great deal that can be before going to high-resolution spectroscopy, however. It is suggested that one start with low-resolution spectroscopy and master that before trying higher resolution work. Part of this is because probably 75 % of the effort is in the software processing. Most of this is the same regardless of resolution. Obtaining low-resolution spectra and processing them is an excellent way to learn spectroscopy. The other part is one can start with low-resolution spectroscopy for a few hundred dollars. Venturing into high-resolution will require several thousand dollars worth of equipment.

Astronomical Spectroscopy Equipment

Telescopes and Mounts

As mentioned earlier almost any telescope can be used, reflector or refractor, ALT/AZ or Polar mounted and a variety of focal lengths. For low-resolution work on bright stars, exposures will be is the sub-second range so tracking is not important. For high-resolution work and where fainter objects are imaged, time exposures will be required. A solid aligned mount is required for high-resolution work. Computerized ALT/AZ telescope will also work fine. The biggest disadvantage of the ALT/AZ mount is a clearance problem when using a LISA or Lhires III or even an ALPY 600 spectrograph. This is especially a problem when observing near the zenith, which is where the best observations are. A Polar mount will allow a large area beneath the telescope when observing near the zenith. For a mount a tripod will work, but if it needs to be set up each observing session, that effort can get old fast. A better approach is some kind of permanent mount. Using a permanent pier and Polar mount, once adjusted, will save many hours and make the observing sessions much more pleasurable.

Naturally the larger the aperture the more light (more photons) is available to be spread out into a spectrum and thus shorter exposures can be used or fainter stars observed. For the equipment listed, a 16″ aperture is about maximum for optimum spectroscopy. Larger instruments can be used, but there is a fast diminishing of returns. The Lhires III is optimized for a 12″ aperture. The Star Analyser is supposed to work best at f/5, but experience shows it works well at f/10 too and probably most focal lengths. The Star Analyser can also be used directly on a telephoto lens of a DSLR camera. The Lhires III works well at f/10 or f/6.3.

At the Hopkins Phoenix Observatory a simple, but permanent roll-off-roof observatory is used. The observatory is 8-foot square and 7-foot high with a fixed pier with polar mount and 12″ LX200 GPS telescope. This observatory is used for high-resolution work with the Lhires III (Figs. 2.1 and 2.2).

Fig. 2.1 Hopkins Phoenix observatory

Astronomical Spectroscopy Equipment

Fig. 2.2 Permanent pier and 12″ LX-200 GPS telescope

A simpler setup was developed for low-resolution work with an 8″ LX90 telescope and Star Analyser. No observatory. The exact dimensions of the following are not critical. A 1-foot deep 2-foot square hole was dug and filled with concrete. A 4-foot long by 12-inch diameter cardboard Sono tube was then inserted into the concrete. A 5-foot long 3-inch diameter steel water pipe, threaded at the upper end was inserted into the center of the Sono tube and also filled with concrete. A 3 inch threaded pipe flange was then screwed onto the top of the pipe. A 10-inch square by quarter inch thick aluminum plate was then screwed to the flange. The LX90 wedge was then attached to the aluminum plate. This has proved to be an excellent arrangement. Once the wedge was aligned the telescope can be installed and removed easily. This arrangement requires little setup time. An extension cord provides AC power to a power strip to power the telescope and computer. If the telescope is to be used for several nights, it is left on the wedge. To protect it from the weather a 20-gallon commercial black plastic garbage bag is installed over it with a rope tying it at the base (Figs. 2.3 and 2.4).

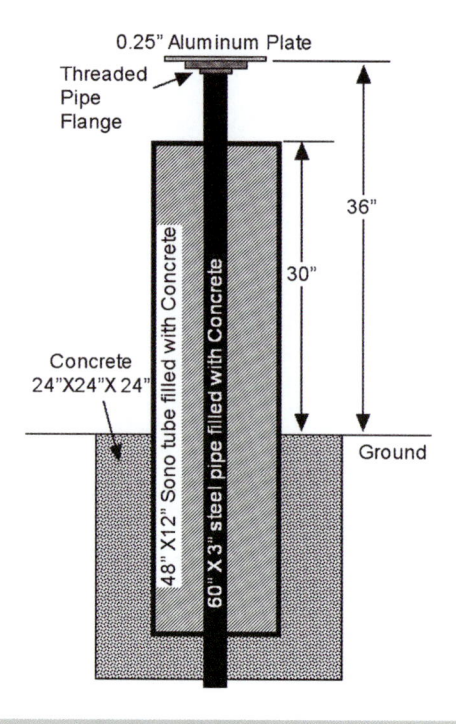

Fig. 2.3 Simple pier for 8″ LX90 and star analyser

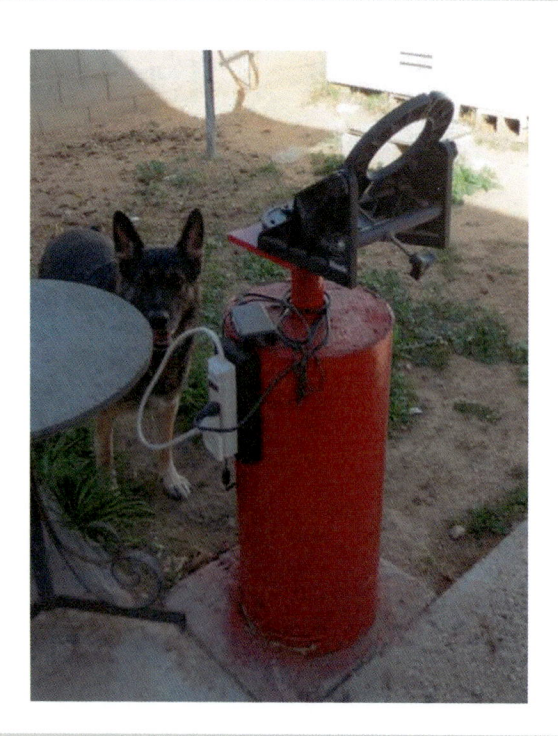

Fig. 2.4 Wedge and pier

Astronomical Spectroscopy Equipment

German Sheppard guard dog Blackie stands guard over the equipment (Fig. 2.5).

Fig. 2.5 8″ LX90 telescope on pier

Digital Cameras

When starting out with low-resolution spectroscopy most any digital camera can be used including DSLR and web cam cameras. A 16-bit monochrome camera will produce an optimum results (best sensitivity, dynamic range and resolution) and is essential for quality high-resolution work. An 8-bit monochrome or 24-bit color camera will work for low-resolution spectroscopy, but they will have a dynamic range of only 256 whereas the 16-bit cameras will have a dynamic range of 65,536.

CCD Versus CMOS

A digital camera has a small chip that is a CCD (charge coupled device) or CMOS (complimentary metal oxide semiconductor) chip. The CMOS chips work well, but the CCD is the preferred type of sensor for most astronomical work. Both CCD and

CMOS work by the photoelectric effect where photons knock off electrons in a cell. The CCD chips are more sensitive in the longer wavelength infrared region. This region is where a lot of work is done on the hydrogen alpha and helium I lines. The CMOS chips are popular in DSLR and other non-astronomical digital cameras. They have a higher data transfer and lower noise that is needed for general camera use. The CCD excels for astronomical use with its better long wavelength response and higher sensitivity to low light conditions.

The CCD and CMOS chips contain rows and columns of pixels in the form of a matrix. Each pixel detects and reports the intensity of photons hitting it. New cameras can have over 1,000 rows and 1,000 columns for total number of pixels greater than one million. When photons hit the pixel, electrons are knocked off and fill a pixel well. The charge in the pixel well is then read and converted to an Analog to Digital Unit (ADU) count. The well is then emptied and ready for the next exposure. Color cameras have 3 or 4 different colored filers over the pixel matrix. For this reason color cameras have much lower resolution than that of a monochrome camera with the same total number of pixels. The filters also reduce the amount of light each pixel receives so the color camera is also less sensitive than a monochrome camera (Fig. 2.6). The following image shows a section of a CCD chip pixel matrix.

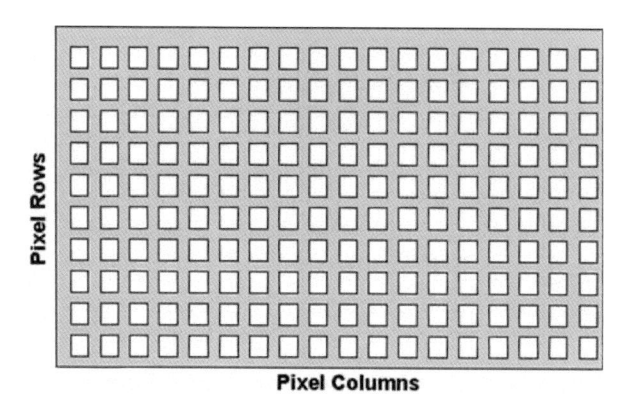

Fig. 2.6 CCD chip pixels

Each of the white squares is a pixel well. For color CCD chips it takes 3 or 4 pixels to produce one color point. Typically the maximum ADU count for a color pixel with a color CCD camera is 256 or 8 bits. Monochrome CCD chips use each pixel for a spot or point. This means the monochrome CCD chip has a resolution of 3 or 4 times that of a color chip of the same size. In addition, monochrome CCD cameras typically are 16 bit cameras and produce pixel ADU counts of from 0 to 65,535 meaning they have a much greater intensity dynamic range.

Astronomical Spectroscopy Equipment

While most any CCD camera can be used with a Star Analyser spectrograph, it is essential that a 16-bit monochrome camera be used for high-resolution work. The Meade DSI Pro II and Orion G3 monochrome cameras both have the same CCD chip. The DSI series is no longer available, but the G3 is available now and has a built in regulated and adjustable TEC cooler. The pictured DSI camera is a DSI Pro II monochrome camera (Fig. 2.7).

DSI Series and Orion StarShoot Cameras

Fig. 2.7 Meade DSI Pro series

One can tell the difference between the DSI Pro and DSI Pro II by the color of the case. The DSI Pro is black and the DSI Pro II is blue. The DSI Pro II also has a bigger CCD chip. The Pro series are monochrome cameras and the DSI without the Pro are color cameras. The Pro series has a sensitivity 230 % and resolution 400 % more that the Color DSI Camera. The Pro series comes with a sliding filter bar. It is best to replace that with a low profile adapter and dark slide, as seen above, available from ScopeStuff (http://scopestuff.com/). The Orion G3 is a better buy now and can be found on sale from Orion for $399 (Fig. 2.8). ATIK also has good cameras, but tend to be more expensive (Fig. 2.8).

Fig. 2.8 Orion StarShoot G3 camera

ATIK 314L+ CCD Camera

Another popular CCD camera with a bigger chip for spectroscopy is the ATIK 314L+. At over $1,600 it is a bit expensive compared to the DSI and Orion series. Another drawback is the larger chip requires a bigger camera and camera space is limited on the Lhires III. Some observers have been able to use them, however. It may require a smaller guide camera so all the cameras can fit. The ATIK 314L+ uses a Sony ICX285AL CCD chip (Fig. 2.9).

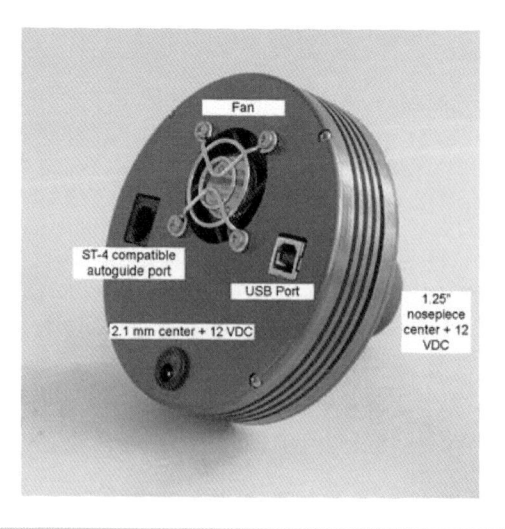

Fig. 2.9 ATIK 314L+ monochrome CCD camera

Table 2.1 shows the specifications for the CCD chips used for the Meade DSI series cameras as well as the Orion StarShoot camera and the ATIK camera.

Table 2.1 CCD chip specifications

CCD specifications
DSI
Color (4 Channel)
Ye, Cy, Mg, G
CCD ICX404AK
1/3″ Chip
510×492 pixels
(250,920 total pixels)
9.6 mm×7.5 mm pixel size
16 Bit ADC
DSI II
Color(4 Channel)
Ye, Cy, Mg, G
CCD ICX419AKL
1/2″ Chip
753×582 pixels
(438,246 total pixels)
8.3 mm×8.6 mm pixel size
16 Bit ADC
DSI Pro
Monochrome
CCD ICX254AL
1/3″ Chip
510×492 pixels
(250,920 total pixels)
9.6 mm×7.5 mm pixel size
16 Bit ADC
DSI Pro II/Orion StarShoot G3 Monochrome
Monochrome
CCD ICX429ALL
1/2″ Chip
753×582 pixels
(438,246 total pixels)
8.3 mm×8.6 mm pixel size
ATIK 324L+
Monochrome
CCD ICX285AL
2/3″ Chip
1,392 h×1,040 v pixels
(1,447,680 total pixels)
6.45 mm×6.45 mm pixel size
16 Bit ADC

Figure 2.10 shows the spectral response curves for the Meade DSI series and Orion StarShoot cameras. Figure 2.11 shows the spectral response curve for the ATIK camera.

CCD Chip Spectral Response Curves

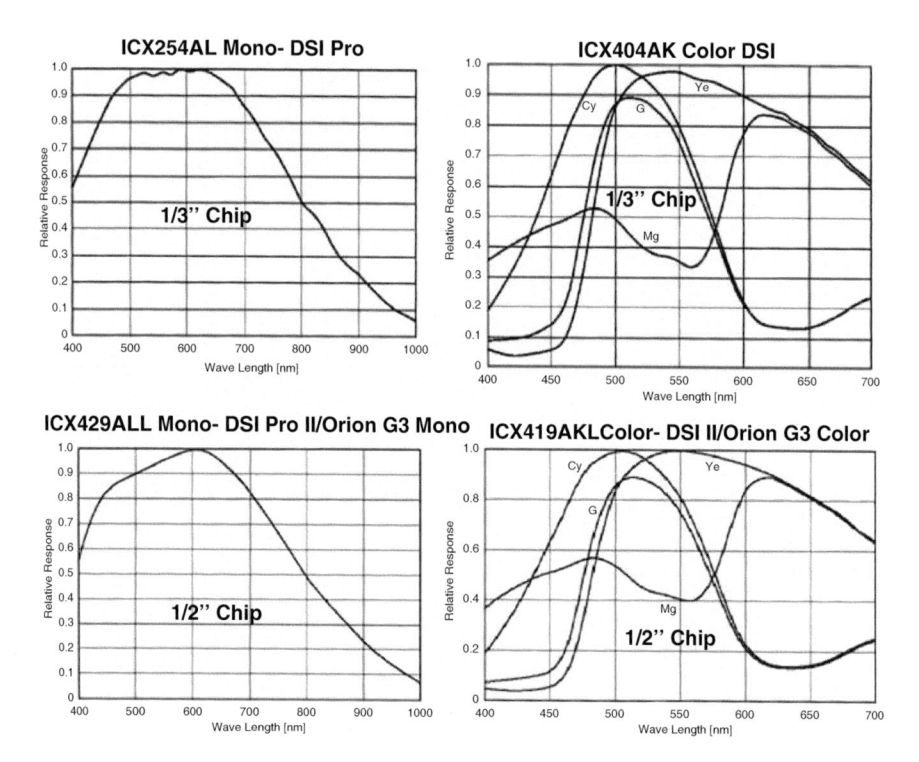

Fig. 2.10 DSI and Orion Star Shoot CCD spectra response curves

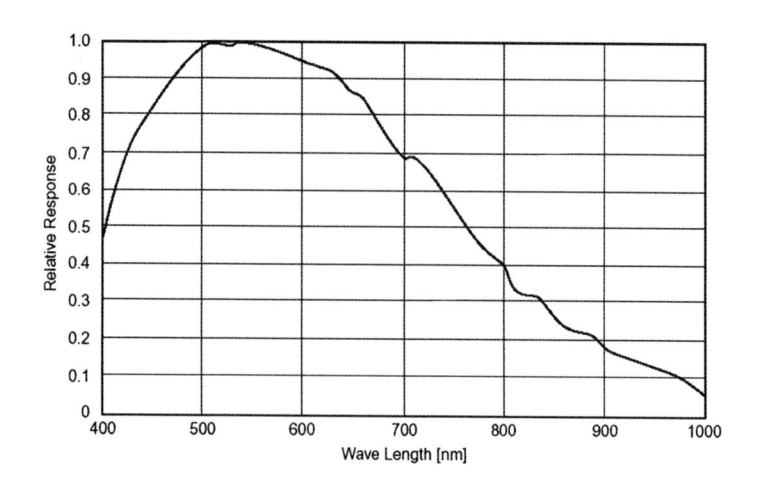

Fig. 2.11 ATIK 314L+ ICX285AL CCD spectra response curve

Spectrographs

Both a prism and/or diffraction grating can be used to produce a spectrum of the visible region of the electromagnetic spectrum. Prisms were used for early spectroscopy, but in order to have sufficient resolution, the prism-based spectrographs tend to get very big and thus usually will only work with large telescopes. Diffraction grating spectrographs can be made much smaller and are more suitable for smaller telescopes. A good example is the Lhires III. The Lhires III uses a Littrow design where the light passes through the same lens twice, first as a collimator and then as a focuser, in a folded optical arrangement.

A diffraction grating is a glass plate or film that has many parallel lines scribed on its surface. The number of lines per millimeter (l/mm) is one factor that determines the possible resolution of the spectrograph. The more line per mm the higher the possible resolution. For the discussed spectrographs the number of lines vary from 100 to 2,400 l/mm. There are two types of diffraction grating, reflection and transmission. In addition the transmission gratings can be used with or without a slit. For those observers in light polluted areas, a slit will essentially eliminate the sky background. Using a slit with a grating enables one to do good astronomical spectroscopy from urban backyards.

In this chapter, the Star Analyser, DIY (Do-It-Yourself), ALPY 600. LISA, Lhires III and eShel spectrographs will be briefly discussed. In Chaps. 3, 4, 5 and 6, the Star Analyser, DIY, ALPY 600, and Lhires III spectrographs will be discussed in detail.

Star Analyser Spectrograph

The Star Analyser spectrograph has a 100 l/mm blazed transmission grating. Note the Rainbow Optics spectrograph is very similar and most of what is discussed regarding the Star Analyser also applies to the Rainbow Optics unit. Both of these spectrographs are used for low-resolution work. Both the Star Analyser and the Rainbow Optics unit have a resolution of 200 or less. The Price of the Star Analyser is under $200 US$ whereas the Rainbow Optics spectrograph is around $250. Both of these spectrographs are normally used in a slitless mode. More information on the Rainbow Optics spectrograph can be found at:
    http://www.starspectroscope.com/
    and the User Manual for the Star Analyser at:
    http://www.patonhawksley.co.uk/staranalyserusermanual.html (Fig. 2.12).

Fig. 2.12 Star analyser spectrograph, credit: Shelyak instruments

DIY Spectrograph

The Do-It-Yourself (DIY) spectrograph has an 1,800 l/mm reflection grating and is fiber optic coupled. The DIY Spectrograph has been available on eBay off and on for $200 US$. This includes everything needed to produce a spectrum line profile except a means of coupling to a telescope to use starlight for the spectrum (Fig. 2.13).

Fig. 2.13 DIY spectrograph

ALPY 600 Spectrograph

In February 2013 Shelyak Instruments announced a new low-resolution spectrograph called the ALPY 600. ALPY is not an acronym, but was conceived in Europe in the Alps. The ALPY 600 uses a grism arrangement. The grism consists of a 600 l/mm transmission grating with a small prism in front of it. This arrangement increases the resolution. The ALPY 600 has a resolution of over 500. The unit can be used with or without a slit. The basic unit has a rotating plate on the front of the spectrograph that allows a slitless mode and several different slit selections. The selectable positions are a 25 mm hole, 50, 100 and 300 mm wide slits plus a 3 mm hole clear position for slitless mode. The unit can be used on a telescope or on a bench via a fiber optic cable. The unit is designed to provide a spectrograph in price and performance between a Star Analyser and LISA spectrograph. The system has three modules. A basic module can be purchased first at minimal cost and other modules added as desired. The other modules consist of a guiding module and wavelength calibration module. The price of the ALPY 600 Is $825 US$ (shipping and exchange rate). All three modules can be purchased for around $2500 US$. The guiding and imaging CCD cameras are not included. User Manuals for the ALPY 600 Modules can be found at (Fig. 2.14):

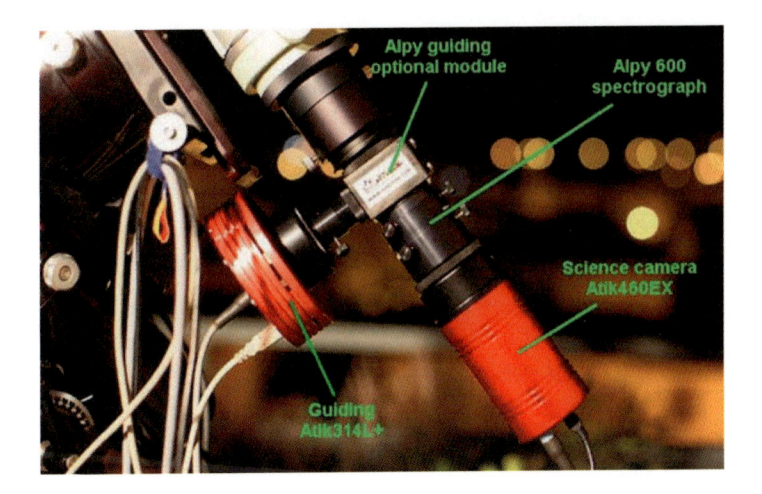

Fig. 2.14 ALPY 600 spectrograph, credit Shelyak instruments

Basic Unit:
http://thizy.free.fr/shelyak/alpy/DC0016A_Doc_Alpy_600_EN.pdf
Guiding Unit:
http://thizy.free.fr/shelyak/alpy/DC0017A_Doc_Alpy_Guiding_EN.pdf
Calibration Unit:
http://thizy.free.fr/shelyak/alpy/DC0018A_Doc_Alpy_Calibration_EN.pdf

LISA Spectrograph

Shelyak also has a low to mid-resolution 600 l/mm reflection grating slit type spectrograph called the LISA, **L**ong **I**maging-slit **S**pectrograph for **A**stronomy. The resolution is between 600 and 1000. The price for the LISA, less CCD cameras and white calibration unit, is around $4,800 US$. The white calibration unit is an additional $500 US$. The LISA User Manuals can be found at (Fig. 2.15):

Fig. 2.15 LISA spectrograph, credit: Shelyak instruments

http://thizy.free.fr/shelyak/Lisa/DC0015A%20LISA%20Pack%20User%20&%20Reference%20Manual%20%28EN%29.pdf
and
http://thizy.free.fr/shelyak/Lisa/DC0012A%20-%20LISA%20User%20Guide%20-%20En.pdf

Lhires III Spectrograph

For mid-resolution and high-resolution spectroscopy a Lhires, **L**ittrow **Hi**gh **Res**olution Spectrograph, III with both 600 and 2,400 l/mm reflection gratings spectrograph will be discussed in detail. The Lhires III is optimized for the 2,400 l/mm grating, but 150, 300, 600 and 1,200 l/mm gratings are available optionally at around $500 USS each. The Lhires III has a resolution of over 17,000 with the 2,400 l/mm grating. The Lhires is available, less CCD cameras for $4,800 US$. This is perhaps the best choice for high-resolution spectroscopy. The LHIRES III Spectrograph User Manual can be found at (Fig. 2.16):

Spectrographs

Fig. 2.16 Lhires III spectrograph, credit: Shelyak instruments

http://thizy.free.fr/shelyak/LhiresIII/DC0004A%20-%20Lhires%20III%20User%20Guide%20-%20English.pdf

eShel Spectrograph

The top of the line spectrograph from Shelyak is the eShel unit. This is an eschell type spectrograph that uses higher order spectra for a wider coverage of the spectrum. This is a fiber optic coupled spectrograph. The price of this spectrograph varies from $15,000 to $25,000, which is way out of reach for most amateurs and thus will not be discussed in detail. The eShel Spectrograph Users Manuals can be found at (Fig. 2.17):

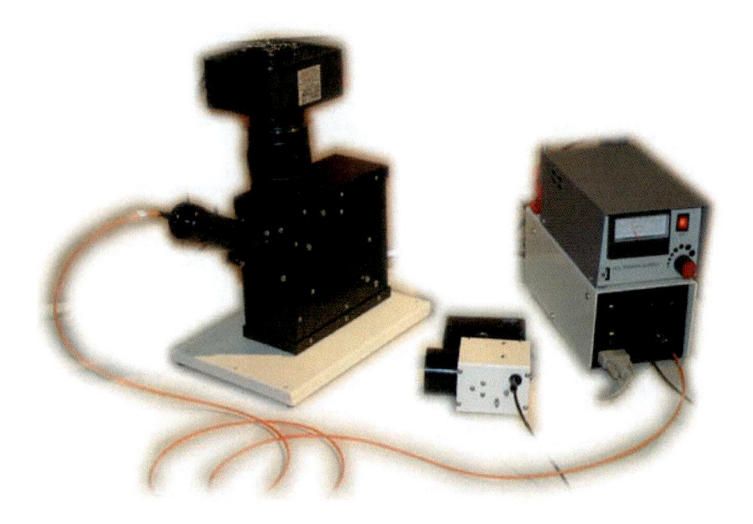

Fig. 2.17 eShel spectrograph, credit: Shelyak instruments

http://thizy.free.fr/shelyak/eshel/DC0010C%20eShel%20Installation%20&%20Maintenance%20Manual.pdf
and
http://thizy.free.fr/shelyak/eshel/DC0009B%20eShel%20User%20Guide.pdf

Taking Spectra

There are some procedural basics that apply to both low and high-resolution spectroscopy. High-resolution usually requires exposures in several minutes, even for bright stars, whereas low-resolution exposures are typically in seconds or milliseconds.

It is suggested that when starting out in spectroscopy that the observer try visual astronomical spectroscopy to get familiar with the spectral orders. This is done with low-resolution spectroscopy and will be discussed with the Star Analyser. Colorful spectra will be seen which can be inspirational.

When a spectrum is created there will be multiple versions of it on each side of the zero order spectrum or star image. Each successive spectrum will be fainter. There will be first order, second order and so on spectra on each side of the zero order spectrum. Normally the first order spectrum should be the spectrum used as it will be the brightest spectrum. For the Star Analyser spectrograph the transmission grating used is blazed in such a way as to make the first order spectrum on one side of the zero order spectrum much brighter than the first order spectrum on the other side. If in doubt which is which, look on both sides of the zero order spectrum

Taking Spectra

and use the brighter one. Rotate the image so the zero order spectrum so to the left and the first order spectrum is to the right. This will produce the proper order of shorter (blue) wavelengths to the left and longer (red) wavelengths to the right. With color cameras this is obvious, but not so with monochrome cameras.

Note: In this Book, most spectra images are monochrome. Some spectrum processing software allows a monochrome spectrum's calibrated line profile to display a synthesized color spectrum based on the line profile wavelength calibration.

Spectral Order

As mentioned earlier and because a diffraction grating will produce multiple spectra, the Star Analyser transmission grating will produce spectra on both sides of the zero order and on just one side with the Lhires III reflection grating, it is important to use only the first order spectrum and in the case of a transmission grating, the brighter first order spectrum. With the Lhires III and 2,400 l/mm grating only the first order spectrum will be seen, however, if other gratings are used, e.g., the 150 l/mm, 300 l/mm, 600 l/mm or 1,200 l/mm gratings, higher order spectra can be seen, but will be dimmer. It is important to use the first order spectrum as it will be the brightest. The following Figure shows the spectrum image at the top with the zero order spectrum in the middle and the two first order spectra on each side with the brightest one to the right. The corresponding line profile of the image is seen below the spectrum image (Fig. 2.18).

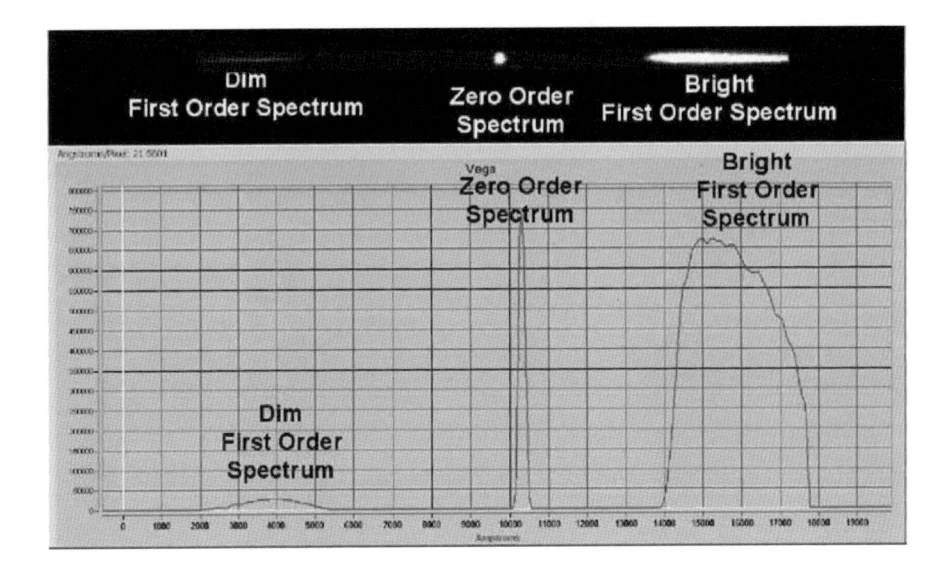

Fig. 2.18 Zero and first order spectra

Exposure

What exposure should you use? There are several factors involved, the brightness of the object, the telescope aperture and the camera sensitivity. There are two points to consider, the saturation point and linearity break point.

Saturation

The saturation point is a function of the dynamic range of the CCD. For a 16-bit CCD the saturation point is 65,535 counts or 2^16 and for 8/24-bit CCD cameras it's 255 counts or 2^8. Each of those actually has one more count, but that is zero. Exposure beyond the saturation point will not produce any more counts. Saturated spectral line profiles will appear with a flat top. While this will not hurt the CCD camera, it should be avoided. Note, the zero order spectrum may be saturated, but because it is not used for any intensity work it is not a consideration.

Linearity

For the best result the exposure should produce a first order spectrum pixel peak ADU count with a high value, but within the linearity range of the CCD. The point where linearity breaks is different for each CCD camera. Typically the break point is around 45,000 ADU counts for a 16-bit camera and 200 for an 8/24-bit camera. To determine a camera's linearity break point a simple experiment can be done. Set up a fixed light source that produces a peak ADU count of 10,000 for a 16-bit camera and 50 for an 8/24-bit camera. Ideally the light source should be producing a continuous spectrum, such as with an incandescent light. Increase the exposure times and note the peak ADU counts. Plot the exposure time vs. ADU counts. The plot should be a straight line (linear) until the break point is reached. At the break point the line will show a knee and further exposure time increase will not produce a corresponding linear ADU count (Fig. 2.19).

Taking Spectra

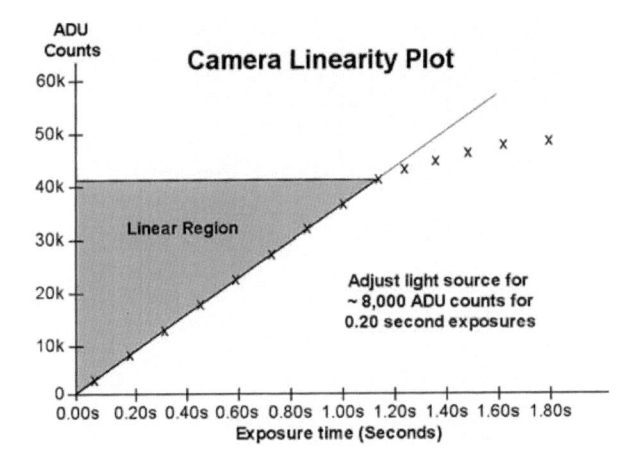

Fig. 2.19 Linearity break determination

Hot Pixels

For exposure times of one second or less, the image should not have any hot pixels. For longer exposures, they can be a problem. A hot pixel is one that has been turned on completely to saturation due to thermal noise and not photons. If the hot pixel occurs within the spectrum image, there will be a large spike in the line profile. These are obvious and can be deleted in the line profile. However, it is better to eliminate as many hot pixels as possible by using a dark frame. To create a dark frame cover the CCD or telescope so no light gets to the CCD chip. Take an image. The exposure should be at approximately the same CCD temperature as the spectrum exposure and the same exposure length. The dark frame is then subtracted from the spectrum image file. The spectrum image is then ready for spectroscopy processing (Fig. 2.20).

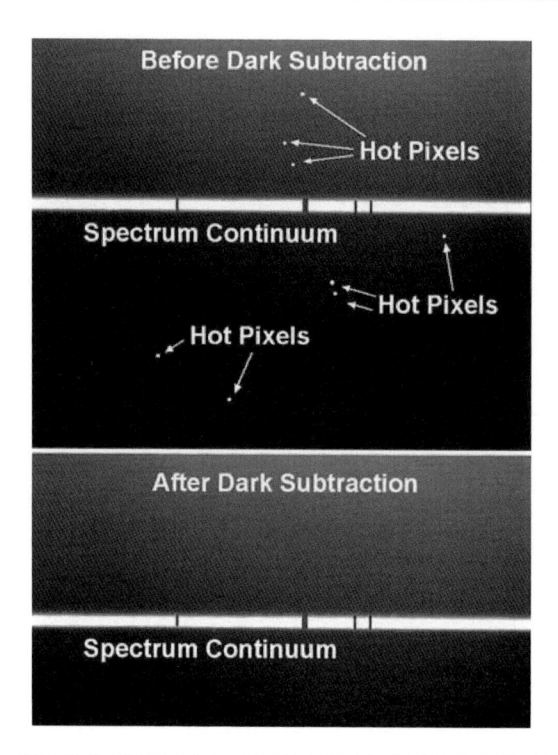

Fig. 2.20 Hot pixels and dark frame subtraction

Pixel Maps

Remember a CCD camera detector is composed of rows and columns of pixels in the form of a matrix. When an exposure is taken, each pixel charge is read and converted to a corresponding ADU count and assigned a matrix coordinate (Y-row and X-column) with that value. A pixel map showing the ADU counts of a specified area of the image can be produced with most imaging software and with the RSpec spectrum processing software. For the first example, the pixel $Y=197$, $X=532$ has an ADU count of 16,433. In the second example a map is not produced, but a read out of the maximum pixel ADU for a selected area of the image can be seen. The important thing here is to adjust the exposure so that the maximum pixel ADU count for the first order spectrum is less that the 65,535 (2^{16} for 16-bit detectors) saturation count and ideally within the linear region of the CCD detector.

Taking Spectra 65

Most image processing software programs allow a means of examining the pixel map. The AutoStar Suite has an Image Processing section that can be used to examine the pixel ADU intensity counts (Figs. 2.21 and 2.22).

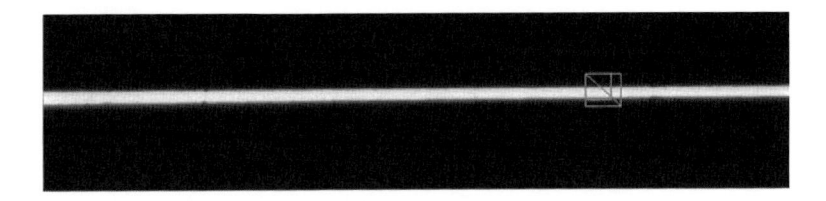

Fig. 2.21 Pixel map area selection with AutoStar image processing

```
Pixel Value Display (528,184) to (552,208)
         528      529      530      531      532      533      53
    184 | 1706.00  1646.00  1701.00  1704.00  1671.00  1793.00  1686
    185 | 1718.00  1723.00  1767.00  1751.00  1770.00  1782.00  1742
    186 | 1784.00  1808.00  1773.00  1819.00  1810.00  1802.00  1779
    187 | 1898.00  1893.00  1844.00  1861.00  1918.00  1950.00  1912
    188 | 1879.00  1967.00  1921.00  1983.00  2028.00  1918.00  1963
    189 | 2140.00  2065.00  2112.00  2133.00  2113.00  2086.00  2117
    190 | 2256.00  2262.00  2241.00  2182.00  2247.00  2166.00  2166
    191 | 2495.00  2480.00  2462.00  2454.00  2429.00  2512.00  2442
    192 | 2762.00  2687.00  2707.00  2724.00  2802.00  2739.00  2839
    193 | 3114.00  3161.00  3149.00  3109.00  3274.00  3182.00  3247
    194 | 3905.00  3734.00  3839.00  3853.00  3715.00  3893.00  3866
    195 | 5023.00  4981.00  5010.00  5167.00  5203.00  5106.00  5134
    196 | 7832.00  7740.00  7877.00  7839.00  7927.00  7842.00  7949
    197 | 15670.00 15611.00 15742.00 16166.00 16433.00 15891.00 16300

         [ Write To Log ]         [ Print ]              [ OK ]
```

Fig. 2.22 AutoStar suite image processing pixel map

Other software also provides means to examine the pixel ADU intensity counts. The spectrum processing program RSpec as an option to show the pixel map and intensity ADU counts for a spectrum image (Fig. 2.23).

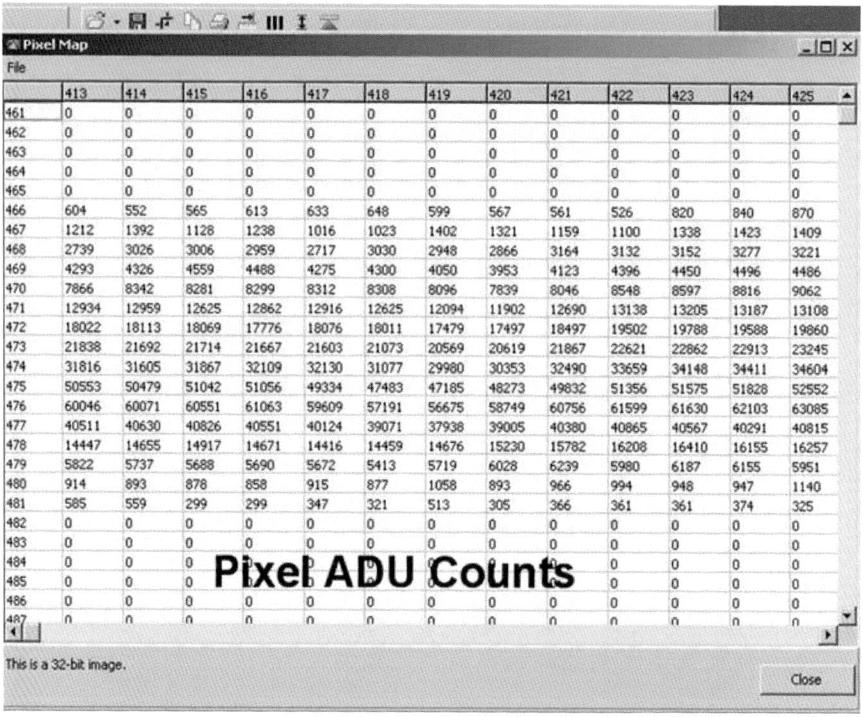

Fig. 2.23 RSpec pixel map

The image processing portion of the Orion Camera Studio software allows examination of the spectrum image when the cursor is positioned (Fig. 2.24).

Taking Spectra

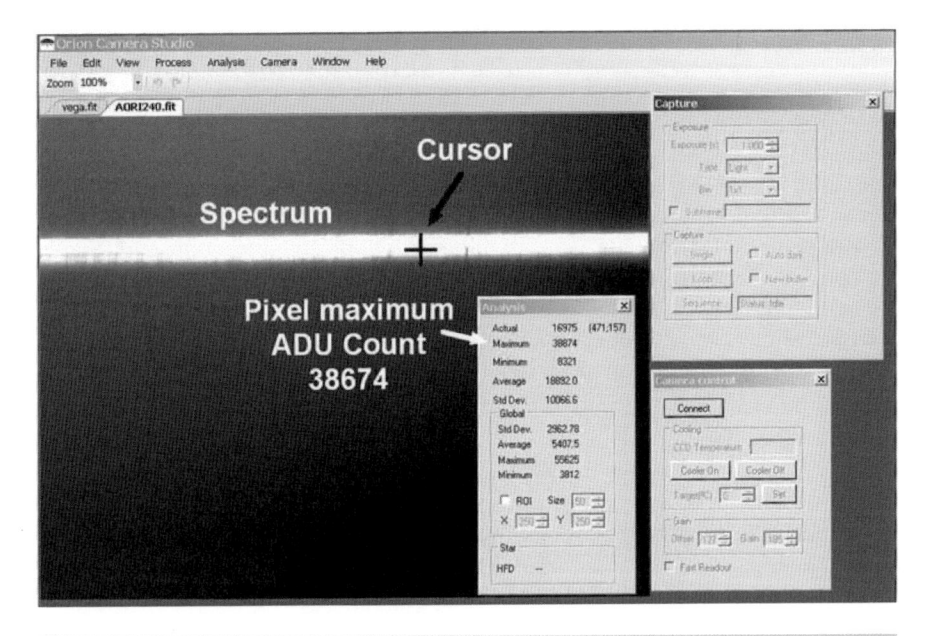

Fig. 2.24 Orion camera studio pixel count

The spectrum processing program VSpec has a means of examining the pixel ADU intensity counts similar to the Orion program (Fig. 2.25).

Fig. 2.25 VSpec pixel count

Dark Frames

A dark frame is an image taken at a specific exposure and temperature with no light falling on the CCD chip. Since no photons are hitting the detector, any pixels that show an intensity ADU count are hot pixels or bias pixels. The hot pixels will appear as bright points in the image. When a spectrum image is taken at the same exposure time and temperature, the same pixels will be hot pixels. The dark frame can be used to subtract those hot pixels (and bias pixels). This can be seen very nicely on long exposures. Before the dark frame subtraction, the image may have multiple bright points. While the hot pixels that are located far from the spectrum are of no concern, any hot pixels lying on or near the spectrum can make the spectrum appear to have an emission line at that point. It is therefore very important to subtract the dark frame. Note, there is also a bias frame, but when subtracting the dark frame, the bias frame is included and thus also subtracted.

With typical low-resolution spectroscopy images that are no longer than 1.0 s, dark frames are usually not needed. Since high-resolution imaging requires exposures in minutes or longer, dark frames should be taken. As noted above, the dark frames should use the same exposure time and approximate temperature of the detector as when the spectrum image is taken. These can be taken before or after the wavelength calibration and spectrum images. The dark frame is then subtracted from the spectrum image prior to the spectrum processing. For the best results, but not always needed, several dark frames of the same exposure and temperature are taken and then averaged. The averaged dark frame is then used. These dark frames can be reused if the CCD temperature and exposure times are the same.

Flat Frames

Flat frames are used to calibrate the CCD or CMOS detector for uniform sensitivity. The sensitivity from pixel to pixel can vary. This is true for both CCD and CMOS detectors. The CCD chips use a common Analog to Digital Converter (ADC) to convert the electric charge to a digital number. CMOS chips have an ADC on each pixel. While the CMOS arrangement provides a faster readout time, it also introduces a variation in the ADC gains from pixel to pixel. Correcting the pixel-to-pixel sensitivity if very important for astrophotography and photometry. It is of lesser importance for spectroscopy. With astrophotography slight imperfections in the optical train and of course the famous dust donuts, faint donut shaped objects in the image, can ruin the image. Flat frames can be used to correct those problems. With photometry, doing a flat frame correction can increase the accuracy of the magnitude determination. The value of flat frame correction for spectroscopy is debatable. Many observers skip using flat fields particularly for low-resolution work.

Because there is a slight variation in the pixel-to-pixel sensitivity of a CCD or CMOS chip, a pixel map that corrects the variation can be used. This will calibrate the chip so all pixels appear to be of uniform sensitivity, i.e., the same light produces the same ADU count for each pixel. Unlike dark frames, flats are much more complex.

Taking Spectra 69

To keep things straight some definitions are in order. First, the flat image taken is called the raw flat image. That image must have a corresponding dark frame subtracted. The resulting image is then the flat field. If the flat field is examined using a pixel map all the pixels will be seen to be about the same ADU count, but the values will vary. What is wanted is a flat calibration frame (Fig. 2.26).

Fig. 2.26 Raw flat image

The following Figure shows a pixel map of the raw flat image where pixel intensity ADU counts vary around 40,000 (Fig. 2.27).

	430	431	432	433	434	435	436	437	438	439	440
432	38097	37967	38174	37791	38306	38242	38254	38128	38280	38052	38411
433	40647	40627	41065	40766	40506	40648	40740	40249	40592	40760	40492
434	37974	37756	37827	37760	37907	38133	38258	38144	38188	38245	38294
435	40648	40612	40574	40386	40618	40543	40638	40627	40535	40148	40380
436	37537	37821	37974	37956	38007	38224	38352	38061	37834	38106	38256
437	40312	40769	40277	40648	40460	40772	40572	40278	40641	40339	40514
438	37872	37544	37452	38138	38046	38082	38265	38103	38080	38537	38084
439	40959	40382	40462	40297	40401	40412	40000	40305	40532	40292	40526
440	37685	37795	37803	37644	37750	38080	38097	37746	37963	37836	38066
441	40775	40499	40653	40923	40610	40816	40824	40350	40365	40098	40549
442	37815	37697	37933	38014	37834	37931	38249	38085	38020	38226	38324
443	40571	40606	40859	40588	40631	40620	40472	40356	40118	40537	40326
444	38201	38009	37789	38233	38202	38083	38007	38004	38283	38588	38223
445	40823	41068	41397	40830	40798	40652	40243	40490	40580	40582	40660
446	38013	38151	38195	38053	38105	38098	38066	38037	37687	37753	38041
447	40429	40592	40747	40349	40410	40314	40690	40138	40257	40065	40305
448	37574	38060	38218	37649	37754	37846	38099	38039	37833	37876	38112
449	40490	40561	40605	40349	40582	40335	40037	40289	40354	40105	40303
450	37822	37935	37459	37765	37678	37720	37968	38064	38016	37801	37782
451	40264	40025	39953	40263	39844	40097	40045	40280	40353	40079	39993
452	37464	37314	37536	37571	38114	38049	37934	37718	37828	37725	37508

Fig. 2.27 Raw flat image pixel map

A calibration flat frame is created by first calculating an average value of all pixel ADU counts. That average number is divided into each pixel's intensity ADU count. This produces a new pixel map with all the ADU counts around 1.00. The following Figure shows the flat frame pixel map. This is a pixel map of the normalized image and is the flat frame that is used to calibrate an image. This is the calibration flat frame. The calibration flat frame is then divided, not subtracted, into the spectrum image. Dividing each of the spectrum image's pixels by a number around 1.00 will make a small correction in the count and adjust the pixel map for uniform pixel sensitivity (Fig. 2.28).

	281	282	283	284	285	286	287
169	1.09	1.08	1.09	1.09	1.09	1.09	1
170	1.00	1.00	1.00	1.00	1.00	1.00	1
171	1.09	1.09	1.08	1.08	1.08	1.07	1
172	1.00	0.99	0.99	0.99	0.99	0.99	0
173	1.08	1.09	1.08	1.08	1.08	1.08	1
174	1.00	0.99	1.00	1.00	1.00	0.99	1
175	1.06	1.08	1.08	1.08	1.08	1.08	1
176	0.99	1.00	0.98	0.99	0.99	0.99	1
177	1.08	1.07	1.08	1.08	1.08	1.07	1
178	0.98	0.99	0.98	0.99	1.00	0.98	1
179	1.08	1.08	1.08	1.08	1.08	1.07	1
180	1.00	0.99	0.99	0.99	0.99	0.99	1
181	1.08	1.09	1.09	1.10	1.08	1.08	1
182	0.99	0.99	1.00	0.98	1.00	1.00	1

Fig. 2.28 Flat frame pixel map

RAW flat frames are taken by uniformly illuminating the CCD chip with a continuous spectrum from an incandescent light. Do not use a fluorescent light. With high-resolution spectroscopy because CCD sensitivity is wavelength dependent, several raw flats should be taken with the grating set for different bands. Take flats every 500 Å from 4,000 Å up to 8,000 Å. These are raw flat images. Exposure times should be used so that the pixel intensity ADU counts are about 50 % of the saturation count. For 16-bit ADC cameras the count should be between 25,000 and 45,000. For 8-bit ADC cameras the counts should be around 125.

The raw flat image must have dark flats of the same exposure time and temperature subtracted from each raw flat image. The results will still not be ready to be used, however. Most imaging programs handle the flats by normalizing the image. This is done by taking the raw flat image pixel ADU count average for the whole chip and dividing that number into each pixel ADU count. This produces a pixel map with ADU counts around a value of 1.00. This is the real calibration flat frame. This map is then divided into the main images (after darks have been subtracted from the main image) to correct all the pixels for varying sensitivity. To take a raw flat and divide it into the main image without the darks or normalizing it is very bad and will produce an image that is severely corrupted.

The value of the extra effort to take and use flats for spectroscopy can be debated. A similar averaging calibration without flats can be done two ways, movement in

Taking Spectra

the slit and binning. The spectrum image is typical a dozen or more pixels in height. This and any movement back or forth in the slit of the star image will increase the number of pixels and height of the spectrum image. Since the line profile sums these column ADU counts, any small variations tend to average out (Figs. 2.29 and 2.30).

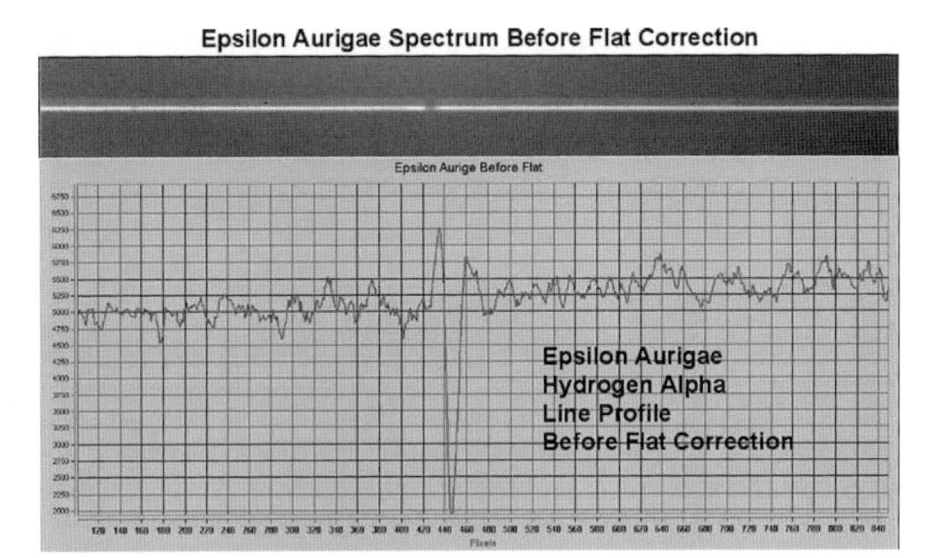

Fig. 2.29 Spectrum and line profile without flat frame correction

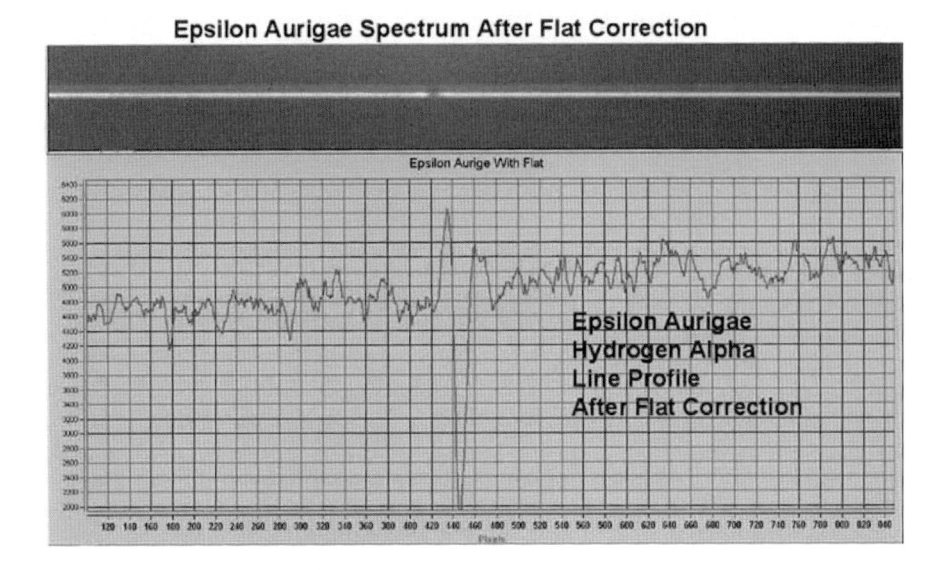

Fig. 2.30 Spectrum and line profile with flat frame correction

Note, some imaging processing programs make this very simple, but be careful that you know for sure what is happening. For example some programs allow you to load the spectrum image and do a calibration by loading a dark frame and flat. The question is which flat, the raw flat image, flat field or calibration flat frame? Usually the flat is the raw flat image. Some imaging programs do all the work for you and you just need to take the raw flat and the program makes and uses the calibration flat. Experiment to make sure you know what is going on.

Image Rotation

As with low-resolution spectroscopy it is also very important that the spectrum image be horizontal for high-resolution work. The orientation of the spectrum should be adjusted by rotating the spectrum imaging camera. With low-resolution work rotating the grating with respect to the camera is needed. If an important image is produced and later found to have a titled spectrum, it may be too late to correct the spectrum at the spectrograph. In such a case most spectrum processing software allows the image to be rotated with software. This is a last resort and while this rotation does work, it can trim the spectrum and produce strange artifacts. Additionally it is very important to make sure the longer (red) wavelengths are to the right in the image. If reversed the spectrum imaging camera should be rotated 180°. It is also possible to rotate the spectrum 180° with software, but best to have the orientation correct when imaging the spectrum. The following Figure shows an extreme of a high-resolution spectrum image tilt (Fig. 2.31).

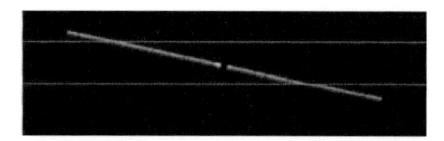

Fig. 2.31 Spectrum image orientation

Background or Sky Subtraction

For spectra of astronomical objects it is important to subtract the background or sky from the image. Most spectrum processing software has this option. A specified number of pixel rows above and below the horizontal spectrum delimiting lines are used to sum pixel columns in those rows and then subtract that number from the sum of the pixel columns delimited by the horizontal lines. This reduces the "floor" of the line profile. Typically 10 pixels above and below are used. It should be noted that the background subtraction should not be used for the neon line calibration. This is for two reasons. First there should be no background or at least an insignificant background. Second the pixels above and below the delimiting lines for the neon spectrum are still the neon spectrum and will result in eliminating or seriously diminishing the neon lines in the line profile.

Horizontal Binning

A spectrum's signal-to-noise can sometimes be improved by using horizontal binning. With the RSpec program this is where adjacent pixels to the right in a row are averaged to create a new pixel value. The degree of binning can be selected. Usually Bin 2, Bin 3, Bin 4, Bin 5 or Bin 6 are available for use. While binning will reduce the wavelength precision with high-resolution images slightly, it may still be of value in reducing the spectrum noise. Low-resolution images with a Star Analyser will show smoother profiles and no significant change in resolution. For the best wavelength calibration with high-resolution images, the minimum or no binning should be used. The following Figure shows an example of Bin 2 horizontal pixel binning. The numbers in the boxes (pixel cells) are intensity ADU counts (Fig. 2.32).

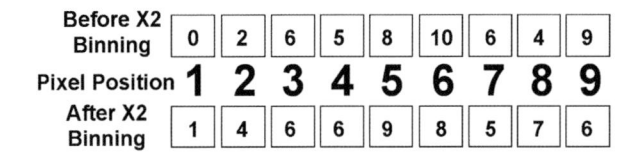

Fig. 2.32 Horizontal binning (Bin 2)

Chapter 3

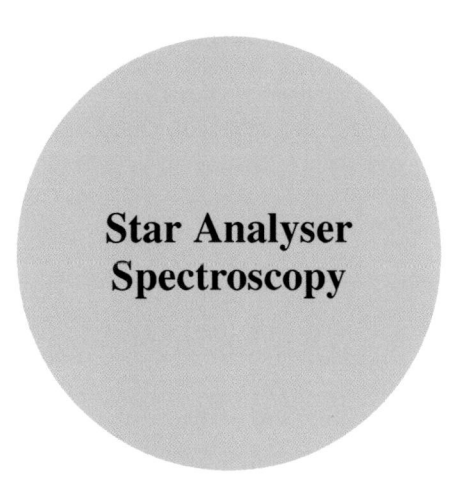

Star Analyser Spectroscopy

Introduction

Astronomical spectroscopy can be very exciting and rewarding, but it can also be very challenging and expensive. For someone just starting out with spectroscopy, the least expensive and least complex system is a simple transmission diffraction grating spectrograph such as the Star Analyser and Rainbow Optics units. No power is needed. No fancy observatory or telescope is needed. The spectrographs define simplicity. Both the Star Analyser and Rainbow Optics spectrographs are two similar inexpensive low-resolution transmission-grating spectrographs. While the Rainbow Optics spectrograph uses a 200 lines/mm grating as opposed to the Star Analyser's 100 lines/mm grating, it costs nearly $100 more (more than 50 % more) without providing a significant increase in resolution. Either one can be used to produce a good low-resolution spectra images. In the following text only the Star Analyser will be discussed in detail, but most everything also applies to the Rainbow Optics spectrograph.

Blazed Gratings

Diffraction gratings can be either a reflective type or transmission type of grating. With the reflective type grating light hits the grating at an angle. The reflected light is broken up into different wavelengths producing a spectrum. A transmission grating is transparent and the light goes through the grating where it is diffracted into different wavelengths producing the spectrum. By nature, the transmission diffraction grating is usually much simpler than the reflection diffraction gratings. Most handheld spectrometers use a transmission diffraction grating. Both the Star Analyser and Rainbow Optics spectrographs also use transmission diffraction gratings, but there is an additional factor that makes their manufacturer more expensive, but provides an enhanced result. They use what is known as a blazed-grating. What is the difference between a non-blazed transmission grating and a blazed one? A non-blazed grating has closely spaced and regular ruled lines on a film or glass substrate that are vertical. With a regular grating the higher order spectra are produced equally in both sides of the grating. The two first order spectra are the brightest and subsequent spectra get dimmer (Fig. 3.1).

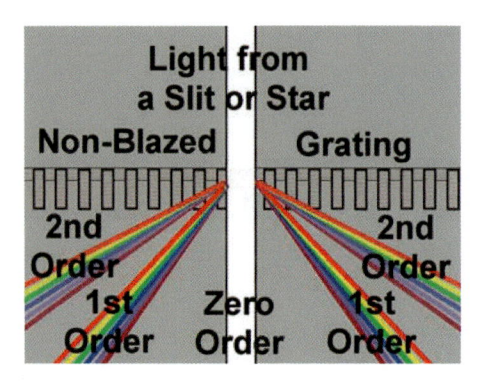

Fig. 3.1 Non-blazed grating

While a non-blazed grating is fine, a blazed-grating has a big advantage. A blazed-grating is one where the lines are not vertical and one side of the line has a sharper angle than the other side. This is kind of a saw-toothed pattern. The saw-tooth pattern causes brighter spectra to be seen on one side of the zero order's spectrum and the corresponding order spectra dimmer on the other side. While any of the higher order spectra, first order on, can be used to create a line profile, the brighter first order spectrum on one side is the one best suited for a good spectrum/line profile and will have a better signal to noise ratio, higher ADU count (Fig. 3.2).

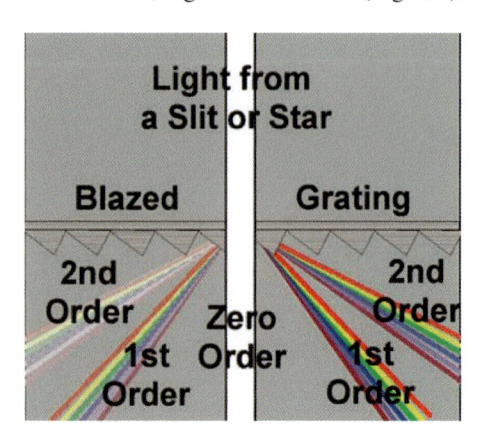

Fig. 3.2 Blazed grating

Normal use of these transmission-grating spectrographs is without a slit. This can be done for astronomical work because star images are very small and in effect mimic a slit. It is even possible to use a slit with these transmission-grating spectrographs, but that complicates the configuration greatly. Using a slit has the big advantage of blocking most of the background light and enabling better selection of a star in a crowded star field.

Star Analyser Equipment

Converging Beam versus Non-Converging Beam Mode

The Star Analyser can be used in either a converging beam or non-converging beam mode. What are these modes? If the Star Analyser is mounted in front of a lens, such as the telephoto lens of a DSLR camera, the light going into the diffraction gratings is from a distant object it is parallel or non-converging. In other words, it is not focused. At first this may seem like it would not work, but the fact is it works well. The diffraction grating produces a spectrum image of the object that is then focused by the camera lens and imaged by the camera's detector. This has some advantages and disadvantages. First, this is easy to do and requires no modifications to any equipment. The only equipment required are the Star Analyser, a DSLR camera and tripod. Theoretically, the non-converging beam can produce a higher resolution spectrum image. The value of this is debatable, however. The Star Analyser is a low-resolution spectrograph and any slight increases in resolution will not likely be of any value. The big downside of the non-converging beam mode is there is no light gathering so only bright objects will produce a spectrum image with a good signal to noise ratio. The converging beam mode is where an objective, either a mirror or lens, focuses the object's image. The focused object image is then impressed on the Star Analyser's diffraction grating producing a spectrum image. The first big advantage of the converging beam mode is because there are more photons available fainter objects can be imaged. For more serious spectroscopy a converging beam mode using a telescope is best. Remember, a spectrum image is just a small part of the original object total number of photons being observed. Since a telescope will gather many more photons fainter objects can be observed and still produce good spectra. Using a converging beam mode from a telescope will produce a brighter image of the object. To use the converging beam mode with a DSLR camera requires the Star Analyser to be placed between the camera and the telescope. Typically astronomical cameras can accommodate a 1.25" nosepiece that has a threaded inside to accept filters. The Star Analyser can be screwed into the nosepiece like a filter and the assembly inserted into the eyepiece holder of the telescope.

Transmission Gratings

As mentioned above, there are at least two commercially available transmission-grating spectrographs that can be used for astronomical spectroscopy, the Star Analyser and Rainbow Optics spectrographs. While either of these units will work fine, the Star Analyser will be discussed in detail (Figs. 3.3 and 3.4).

Fig. 3.3 Star analyser

Fig. 3.4 Rainbow optics

Star Analyser

The Star Analyser is a slitless transmission diffraction grating spectrograph with a blazed-grating that can be used in either a converging light beam mode (with a telescope) or non-converging light beam mode on the front of a lens such as with a telephoto lens on a DSLR or other digital camera.

No matter which mode it used it is important to get the Star Analyser's the spectrum image as horizontal as possible. The spatial rotational relationship between the Star Analyser's lines in the diffraction grating and the detector's alignment of its pixel rows and columns determines the rotational position of the spectrum image. It is also important to get the rotation of the spectrum image such that the shorter wavelengths (blue region) are to the left and the longer wavelength (red region) to the right. A typical high-dispersion spectrum image (25–30 Å/pixel) will include the zero order's spectrum and if that is centered, the two first order spectra will also be seen. The region of each first order spectra closest to the zero order's spectrum will be the shorter or blue wavelengths. This means the first order's spectrum to the right is the one that should be used. If it is not the brighter of the two first order spectra, then the Star Analyser should be rotated 180° to put the brightest first order's spectrum to the right of the zero order's spectrum. While the rotation can be adjusted in software, it is better to get it correct at the spectrograph before taking the image. This can be a frustrating endeavor, but it is important. Plan to take some time to get it right. It will require removing the Star Analyser and adjusting its rotational position and then re-installing it and checking the spectrum image's position. It is important to screw the locking ring (provided with the Star Analyser) into the port before adding the Star Analyser. The locking ring should be rotated along with the Star Analyser. Once the proper position has been found, the locking ring can be used as a stop. An orientation mark is provided on the Star Analyser for reference. A small flat bladed screwdriver in the notch can be used to turn the locking ring. For that reason, when installing the locking ring, make sure the notch is in the "out" position. The fit of the locking ring is usually loose enough, however, that a finger can usually be used to do the adjustment. The rotational position of the Star Analyser relative to the detector's orientation is very sensitive. Once the orientation is close it may take several small adjustments to fine tune the orientation. Sometimes people do not find the locking ring in the box with the Star Analyser. Usually there is a foam pad in the box. The locking ring can be found under the pad. It is small and easy to miss. As will be discussed below, there is a way using different techniques to avoid all the frustration with getting the spectrum image horizontal (Fig. 3.5).

Fig. 3.5 Star analyser and locking ring

In addition to the Star Analyser, some kind of a detector is needed to register the spectrum. For visual work with the Star Analyser using an eyepiece connected to a telescope, the eye is the detector. Most any digital camera can be used as a start. As noted earlier video cameras, web cams and DSLR cameras have produced good spectra. Again, for the best results a 16 bit monochrome CCD camera is suggested as the detector.

Grism

A 3.8° prism was introduced in the summer of 2013 that can be used either in front of or behind the Star Analyser to form a grim. The addition of the prism will slightly improve the resolution and reduces the coma that is sometimes seen in the red region when using the Star Analyser. The prism uses standard 1.25″ threads and can be screwed directly to the Star Analyser. There is an alignment mark on the prism holder that shows the correct orientation with the Star Analyser. There is another (supplied with the prism) locking ring that can be used to lock the prism in place. When used in a grism configuration the spectrum produced by the Star Analyser becomes significantly non-linear. Because of this a non-linear wavelength calibration is required to create an accurate wavelength calibration of the spectrum's line profile. This means three or more lines must be used for the calibration. When mounted behind the Star Analyser the prism adds 15 mm to the spacing distance to the detector (Fig. 3.6). The prism can be purchased from Woodland Hills (Model#: OP0040) for $97.50. See http://telescopes.net/store/manufacturer/shelyak-instruments?

Star Analyser Equipment

Fig. 3.6 Star analyser prism/grism

Digital Cameras

Most DSLR cameras use CMOS detectors rather than CCDs. That in itself is not bad, but perhaps the biggest problem is few if any DSLR cameras will produce 16 bit monochrome images with a .fits format. This means DSLR and other color cameras including video cameras and web cams can be used, but at significantly reduced resolution and sensitivity. In addition the images produced are not in a .fits format. The .fits format is a very important consideration for serious astronomical spectroscopy. Beginners sometimes prefer the color cameras because they produce a color spectrum as opposed to the monochrome spectrum. While the color spectrum is pretty and may be inspiring, the color is of little value. If one really wants to see a color spectrum, once the line profile has been wavelength calibrated, an accurate pseudo color spectrum can be created with software using the wavelength and intensity of the line profile. The monochrome image on the other hand, while not as pretty, usually is a much higher resolution image with much better sensitivity and thus a much higher ADU count. This will produce a better signal to noise ratio for the spectrum image. The bottom line is when starting out, use whatever camera you have. Produce some spectra and process them. Once familiar with the techniques and if the observer desires to do serious spectroscopy.

DSLR Camera

There are at least three ways a DSLR camera can be used with a Star Analyser. The first way is a stand-alone method. Imagine doing astronomical spectroscopy without an observatory or even a telescope. This can be done using just a DSLR camera with a Star Analyser all set on a tripod. This may sound like a great idea and the parallel beam of the non-converging light of the star can theoretically produce higher resolution spectra, but there are some significant downsides to using a DSLR and Star Analyser. First, a DSLR is not designed to take faint astronomical images. Even without the Star Analyser attached, looking through the camera's finder and trying to find and focus on an object, other than the Moon or a very bright star can be extremely frustrating. Once the Star Analyser is added, everything gets dimmer and blurred and the problem is compounded. Because the field-of-view with the camera will be large, sometimes just aiming in about the right direction will work as long as the focus is good. DSLR cameras are color and probably 24-bit which means the dynamic range may be only 256 (8 bits for each channel) for the RGB pixels. For bright objects (1st magnitude and brighter) spectra can be taken with sub-second exposures without tracking. One trick for fainter objects is to take longer exposure, several seconds, but align the camera such that the motion of the star or other object causes the spectrum to stretch vertically as the image drifts. Unlike astronomical images, this actually can help. When processing the spectrum the line profile is the desired results and with the spectrum stretched vertically the column pixels are summed producing a higher ADU count for that column. The downside is except for bright objects in a non-crowed star field, multiple star spectra may be mixed and the spectrum of interest difficult or impossible to separate.

A second method is a more complex arrangement, but one that allows easier imaging and better time exposures. This method is accomplished by using the DSLR camera with Star Analyser as in the first method, but now piggybacked on a tracking telescope. The telescope can be used both to find fainter objects and to track them for a time exposure.

The third method is to use the DSLR camera without the camera lens. An adapter must be used to allow the DSLR to be mounted to the eyepiece port on a telescope. This method uses a converging beam configuration. There are many types of adapters due to the variety of different camera threads. ScopeStuff has a good selection. See http://www.scopestuff.com/ss_wca1.htm. The Star Analyser can then be screwed into the nosepiece of the adapter. This solves the problem of finding and tracking an object as well as providing more photons for the spectrum and thus shorter exposures and/or spectrum images of fainter objects.

If you have a DSLR camera and telephoto lens you can certainly experiment, but for serious work a dedicated 16-bit monochrome CCD camera is needed. Top end astronomical CCD cameras costing many thousands of dollars are not needed.

The Figure below shows the Star Analyser in a non-converging mode using a telephoto lens and DSLR camera. The in the image the Star Analyser is screwed into a separate cell holder, but this is not required (Fig. 3.7).

Star Analyser Equipment

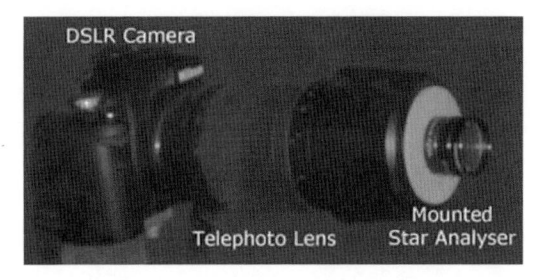

Fig. 3.7 Star analyser with a DSLR camera (non-converging mode)

Non-Converging Light Beam

As shown in Fig. 3.7, to use the Star Analyser in a non-converging light beam it can be mounted on the front of a telephoto lens. The focal length for the telephoto lens is not critical, but should be at least 200 mm. Adapters that allow the Star Analyser to be mounted on a telephoto lens can be purchased from RSpec for $38 plus shipping or a simple one made out of plastic or even wood. A lens cap can also be easily modified for this. See http://www.rspec-astro.com/Store/. The camera telephoto lens can be coupled either to a CCD camera or to a DSLR camera. Aside from the simplicity, this arrangement also makes adjust the Star Analyser's rotational position much easier. The Star Analyser can be easily rotated with respect to the DSLR camera. This arrangement can be used on a tripod for the simplest setup of mounted piggyback on a telescope.

Converging Light Beam Mode

The best use of the Star Analyser is with a converging light beam such as at the focus using a telescope. Using the telescope the spectrum image can be viewed either visually or with a digital camera. When using the Star Analyzer visually, an eyepiece can be used. Most eyepieces have threaded barrels used to accept filters. The Star Analyser is just screwed into the eyepiece barrel like a filter. To adjust the spectrum image's orientation, the whole eyepiece with Star Analyser can be rotated. The magnification or dispersion of the spectrum is determined by the focal lengths of the telescope and eyepiece and the distance between the grating and the eyepiece. With an F/10 8″ telescope and 25 mm eyepiece with the Star Analyser mounted, the zero order and at least the first order spectra should be visible. To show more higher order spectra a 32 mm or 40 mm eyepiece can be used. To expand the image so just the first order spectrum is seen and fill most of the view, a 17 mm or even 12 mm will provide an excellent expanded view of just the first order spectrum. One advantage of starting with an eyepiece is that it can show the higher order spectra and which side is favored by the grating blazing and as mentioned above it is easy to get the spectrum oriented properly because the whole eyepiece can be rotated until the spectrum image is horizontal and with the blue end to the left. Note the locking ring is not needed for this method (Fig. 3.8).

Fig. 3.8 Star analyser with an eyepiece

For use with an electronic detector even simple web cams can be used to experiment with. The ToUcam was a very popular inexpensive color camera that was used when astrophotography with digital cameras took off. In order to use the ToUcam a special adapter must be used in place of the lens that comes with the ToUcam (Fig. 3.9). These adapters can be purchased from ScopeStuff for $19 to $29 (WCA1, WCA2 and FR1A) depending on the telescope used. See http://www.scopestuff.com/ss_wca1.htm.

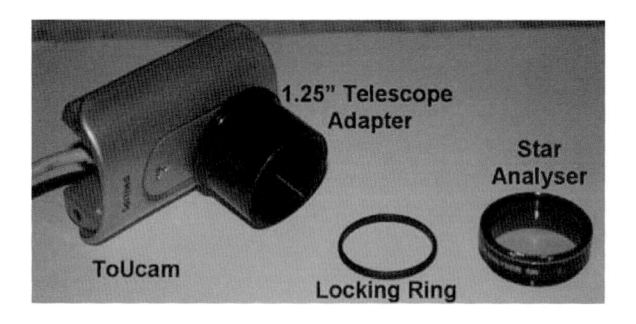

Fig. 3.9 Star analyser, locking ring and web cam

Star Analyser Equipment

The Star Analyser can be screwed into the nosepiece adapter of a web cam. The above web cam is a ToUcam with a 1.25″ adapter. The black locking ring can be screwed in first to lock the position of the Star Analyser relative to the detector. Most CCD cameras have a nosepiece that the Star Analyser can be screwed into (Fig. 3.10).

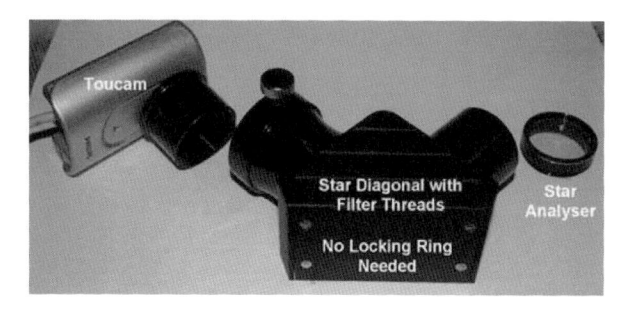

Fig. 3.10 Star analyser, web cam and star diagonal

To increase the size of the spectrum (decrease the dispersion Å/pixel) the distance between the Star Analyser and detector can be increased using special spacers. Something that works surprisingly well and probably will not cost anything is the use of a star diagonal. Spacers allow multiple spacing to be experimented with and can be purchased from RSpec for $10. See http://www.rspec-astro.com/Store/. Be aware that when using a star diagonal, not all star diagonals accept threaded filters or the Star Analyser. It was discovered that a star diagonal worked well when using a ToUcam or DSI Pro series CCD cameras. With the DSI Pro II and a star diagonal typical dispersion is 5–6 Å/pixel. With this arrangement the zero order's spectrum will not be seen, but the whole expanded visible spectrum will be imaged (Fig. 3.11).

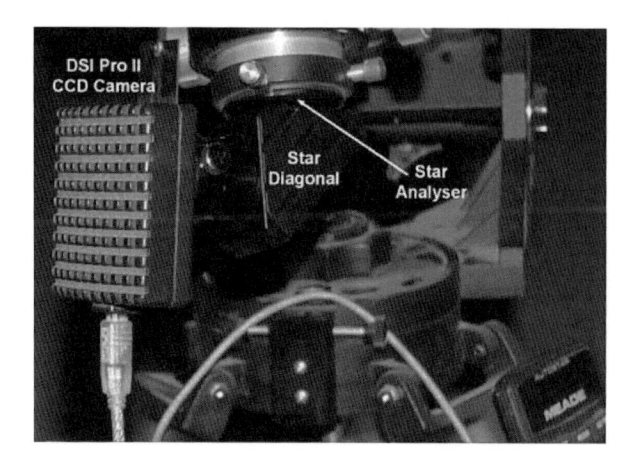

Fig. 3.11 Star analyser, monochrome CCD camera and star diagonal

Telescopes

Most any telescope can be used with the Star Analyser. For less than bright objects, the bigger the objective of the telescope, the better. There is some debate on what focal length is best. Supposedly the Star Analyser works best with an F/5 optical system. Experience has shown that an F/10 system works well and while subjective, the resulting line profiles look better with the F/10 system than an F/6.3 system (close to F/5).

The following Figures show line profiles of a 0.022 s exposure of Betelgeuse using a Star Analyser with DSI Pro II CCD camera. The telescope used was an 8″ LX90 (F/10) first with an F/6.3 reducer to effect an F/6.3 system and then without the focal reducer at F/10. The Figure shows more pronounced features with the F/10 system. So which is better, a faster optical system or one of higher magnification? You decide (Fig. 3.12).

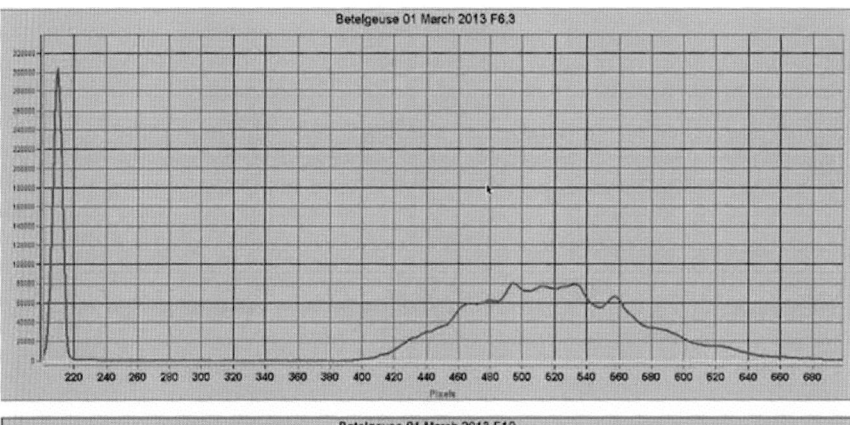

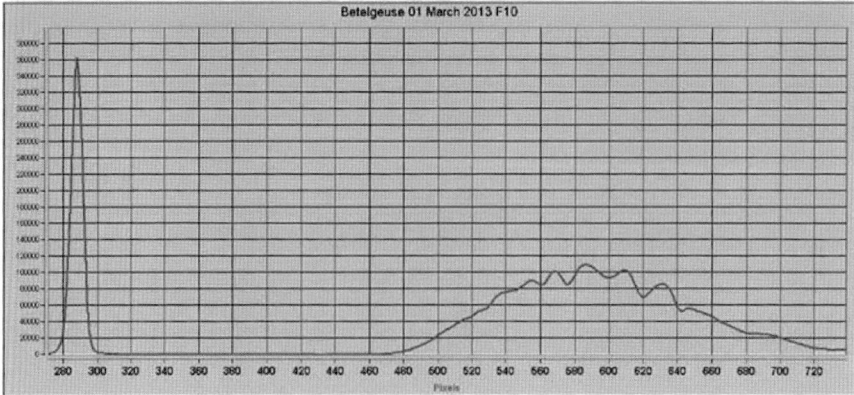

Fig. 3.12 Zero order spectrum profile F/6.3 (*top*) versus F/10 (*bottom*)

Star Analyser Equipment

The bottom Figure with the F/10 exposure shows more pronounced features in the spectrum's line profile than the top Figure at F/6.3 (Fig. 3.13).

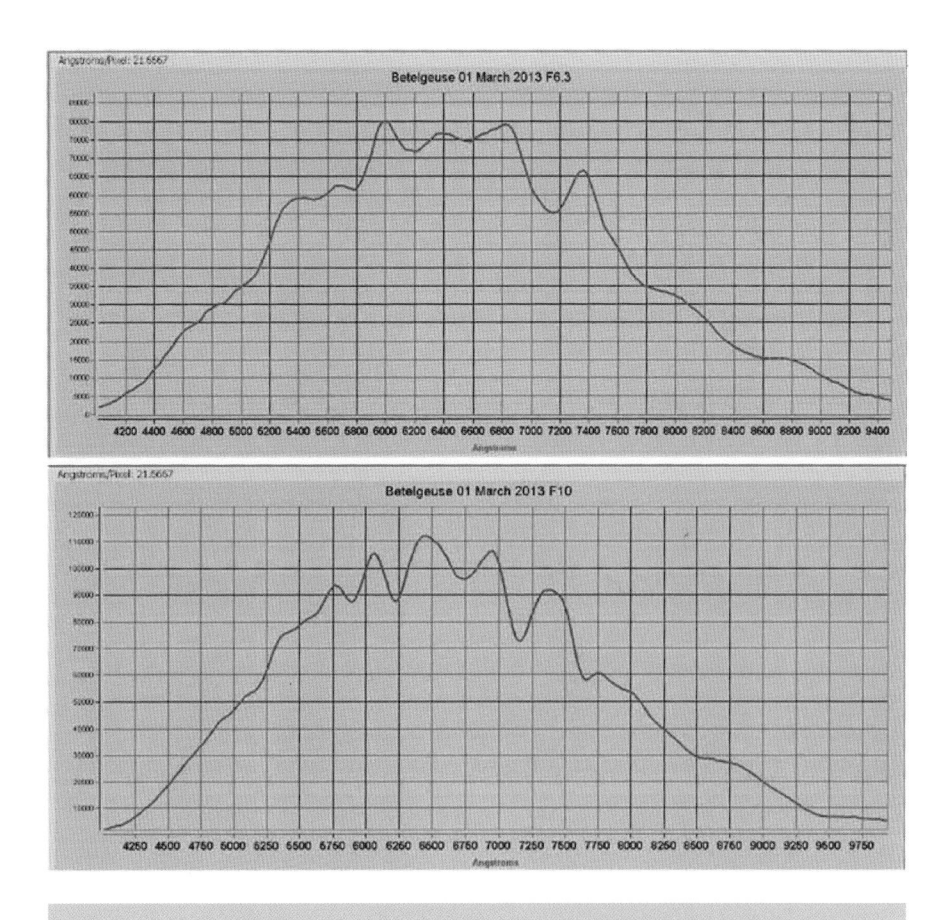

Fig. 3.13 Expanded spectrum profile F/6.3 (*top*) versus F/10 (*bottom*)

The above Figures show expanded line profiles. The more pronounced features of the F/10 exposure can be clearly seen. What is the optimum focal ratio to use? This is subject to debate and much a personal preference. It is probably best to use the basic focal ratio of the telescope you have. Experimenting with focal reducers and Barlows lens' (to increase the focal ratio) is something that can be done.

Taking the Spectra with a Star Analyser

Focusing

When using a DSLR camera and Star Analyser in non-converging mode, the camera should be focused at infinity, at least to begin with. If that does not produce a good focused spectrum image then try focusing on the spectrum image. This can be very difficult as it is hard to tell when the spectrum image is optimally focused. The next best thing is to focus on the zero order's spectrum image, which is the star's image.

When using a converging beam mode with a telescope, if the telescope's focus is too far off, even bright stars will not be easily seen when the Star Analyser is installed. It is important to get the focus approximately set before adding the Star Analyser. The procedure is to first, remove the Star Analyser and using the same optical configuration as will be used when the Star Analyser is installed, find a bright object and focus on it. Once the focus is close, then add the Star Analyser. While the focus will be close it will still need adjusting to produce a sharp spectrum image. Rather than trying to focus on the spectrum, the initial focus should be on the star of interest. This star's image will be the zero order's spectrum image for the star. Once everything appears to be good, then focus on the spectrum. Note that for precise focusing of the spectrum if one end of the spectrum is precisely focused, the other end will be slightly out of focus. This is because of the different wavelengths at the red and blue ends of the spectrum having different focuses. The red end will have a slightly different focus than the blue end. For this reason it may be best to try to just focus on the center of the spectrum or the red end as that is the area most likely to be of interest. Depending on the star, focusing can be very challenging, but interestingly even a less than perfect focus can produce a good spectrum and resulting line profile.

Finding the Object

Finding the object to take a spectrum of can be challenging. More than one astronomer has at sometime taken a spectrum on the wrong object. A DSLR camera may seem convenient to use, but as mentioned above it has some serious drawbacks. First, the camera is not designed for astronomical imaging. The camera's finder will be of little value finding any but the brightest objects. But, because the field-of-view of a DSLR camera, even with a telephoto lens installed, is large, just aiming the camera in the approximate direction of the object of interest may work. The big advantage of this method is just a tripod is needed for the camera. A more complex, but better approach is to mount the camera piggyback on a telescope. The piggyback mount must be such that the camera can be moved relative to the telescope to allow it to be aligned with the telescope. Use a bright object, such as the Moon or bright star, and center the object in the telescope's field-of-view. A 12 mm dual crosshair eyepiece is good to use for getting the object centered. Next adjust the DSLR camera's position on the mount so that the object is also in the center of the camera's finder. Now objects of interest can be found and centered using the telescope and thus the object will also be centered in the camera. A well-aligned telescope with a clock drive or computer controlled tracking can also allow the camera to track a faint object for a time exposure.

As mentioned earlier the Star Analyser works best when used with a telescope in a converging beam mode, but even with the telescope, finding the object can sometimes be a challenge. It is usually easier to find and focus the object of interest without the Star Analyser installed. Then install the Star Analyser being careful not to move the telescope. The addition of the Star Analyser will require refocusing, but the focus should be close.

Spectral Order with the Star Analyser

Viewing the proper spectrum in the spectrum image is very important. Make sure you are viewing the brightest first order spectrum. Remember, there will be two first order spectra, one on each side of the zero order's spectrum. The brightest spectrum will be one of the two first order spectra. Orient the spectrum image so the brightest first order spectrum is to the right and the zero order's spectrum is to the left. The zero order's spectrum is at zero Å and the blue end of the visible spectrum will be the end closest to the zero order's spectrum. The spectrum wavelength gets longer the further it is from the zero order spectrum.

The following Figure shows an image with the zero order and two first order spectra. It is easily seen that one of the first order spectra is much brighter than the other (Fig. 3.14).

Fig. 3.14 Bright first order spectrum

Spectrum Image Orientation

As mentioned above the brightest first order spectrum should have the shorter (blue end) to the left. The blue end is nearest the zero order's spectrum. Make sure the spectrum is as horizontal as possible. To adjust the orientation it will be necessary to rotate the Star Analyser with respect to the CCD chip's orientation. If the Star Analyser is installed in the nosepiece of the camera the Star Analyser must be rotated relative to the camera. This can be challenging, as it requires removing the camera each time the Star Analyser is rotated and then replacing it and checking the spectrum orientation. The locking ring supplied with the Star Analyser can be inserted into the nosepiece ahead of the Star Analyser. By adjusting the locking ring, the Star Analyser's rotation can be stopped at the proper orientation to produce a horizontal spectrum. Once properly adjusted further changes of orientation should not be needed.

If a spacer or star diagonal is used to provided an enlarged image of just the first order spectrum, it is possible to change the relationship between the Star Analyser's grating orientation and detector to get a horizontal spectrum by just rotating the camera relative to the spacer or star diagonal with the Star Analyser. This is much easier that trying to rotate the Star Analyser in the nose of a CCD camera.

Exposure

As noted earlier the proper exposure for the object of interest and optical configuration should be determined. This will require taking multiple images of the spectrum at different exposures and examining the spectrum's pixel ADU intensity counts. For bright stars, e.g., Vega, sub second exposures are usually sufficient. It is easy to over expose the bright stars.

There seems to be some observers who started with astrophotography and want to carry over their techniques from that into spectroscopy. Perhaps the biggest technique is stacking. With astrophotography stacking can be very useful for enhancing the image. The value of stacking images is only true for very bright objects that require short exposure times. If exposure times are in the milliseconds, then it is easy to take dozens or even hundreds of images. The problem with doing this for spectroscopy is that it does not help. The whole purpose of spectroscopy is to produce a good spectrum. Surprisingly, a less than perfect image of a star will produce a perfectly good spectrum. Will stacking hurt? No, but it will not be of much value either.

The best approach to getting a good spectrum image is to experiment. Determine your camera's linearity break, the exposure time that produces the maximum pixel ADU while still in the linear region. For 16-bit CCD cameras the saturation ADU count is 65,535 and typical linearity break occurs around 40,000 ADU counts. Knowing that ADU count, try different exposures for the star. Do not worry about a spectrum yet. Examine the spectrum image and determine the maximum pixel ADU counts in the brightest parts of the spectrum. Adjust the exposure until the counts are at or just below the linearity break count. Use that exposure for the acquisition of the spectrum. For short exposures, a few seconds or sub-seconds and if the atmosphere is very unsteady, take several images and use the one that looks best. For long time exposures, many seconds, minutes or hours, only one exposure is needed. The long exposure will average out any atmospheric variations. Again, the main goal is to get starlight into the diffraction grating to produce a spectrum. Even spectrum images that look less than great can still produce a surprisingly good line profile.

Image Processing

Dark frames

Unless the spectral images are long exposures, more than a few seconds, dark frame calibration should not be needed. When dark frames are needed, most astrophotography type CCD cameras have software that will help acquire and use dark frames. This is another plus for using a dedicated CCD camera as opposed to a DSLR camera. For more information see Sect. 2.3.

Flat Frames

Normally flat frames are of little value with low-resolution images. For more information see Sect. 2.3.

Image Rotation

As mentioned earlier, it is important to have a horizontal spectrum image of the brighter first order's spectrum with the red part of the spectrum to the right. If the zero order's spectrum is in the view it must be to the left. If the final image is not as horizontal as desired, most spectrum processing software allows software rotation of the image. This can sometimes cause problems, e.g., truncating the spectrum ends and strange artifacts. It is best to have the spectrum image sufficiently horizontal without the software rotation. More information on the software rotation will be discussed in the Software Chapter. Again, in addition to having the spectrum horizontal it is important to have the shorter wavelengths (blue) to the left and longer wavelengths (red) to the right. Just make sure the first order spectrum is the brightest one and the zero order star image is to the left. This is very important!

Background Subtraction

The spectrum processing software should allow the background around the spectrum's image to be subtracted. This will be discussed in detail in the Software Chapter.

Low-Resolution Spectrum Processing

Line Profile Creation

To be useful the spectrum image must be used to create a line profile of the spectrum. This along with wavelength calibration and instrument response correction will be discussed in detail in Chap. 7.

Low-Resolution Wavelength Calibrated Star Profiles

Calibrating a line profile requires some detective work. The trick is to try and identify at least two lines. Use those to calibrate the line profile. Then check the calibration by imposing elemental lines from a library. Most spectrum processing software allows that. The hydrogen Balmer lines are always a good check. To help figure what lines are what Richard Walker has produced a free Spectroscopy Atlas of numerous stars as a .pdf that can be downloaded at: http://www.ursusmajor.ch/downloads/spectroscopic-atlas-4.0.pdf

The following wavelength calibrated line profiles of selected bright stars are provided as references to help determine the wavelength calibration of lines profiles of these stars taken by other observers. The identified Balmer lines should provide guideposts for the calibration. Remember once you do the calibration, use the Elements option and Hydrogen Balmer Lines to check how well the Element lines line up with the profile's lines.

Solar Spectrum Line Profile

The following is a spectrum line profile of the Sun. Because the Sun is so bright, it normally cannot be imaged directly. One way is to focus on a small reflection of the Sun. Some people have used a needle and imaged the Sun's reflection off of it. The following Figure is a line profile of a solar spectrum image taken with the DIY spectrograph by just pointing the fiber optic cable toward the outside in the middle of the day (Fig. 3.15).

Low-Resolution Spectrum Processing 93

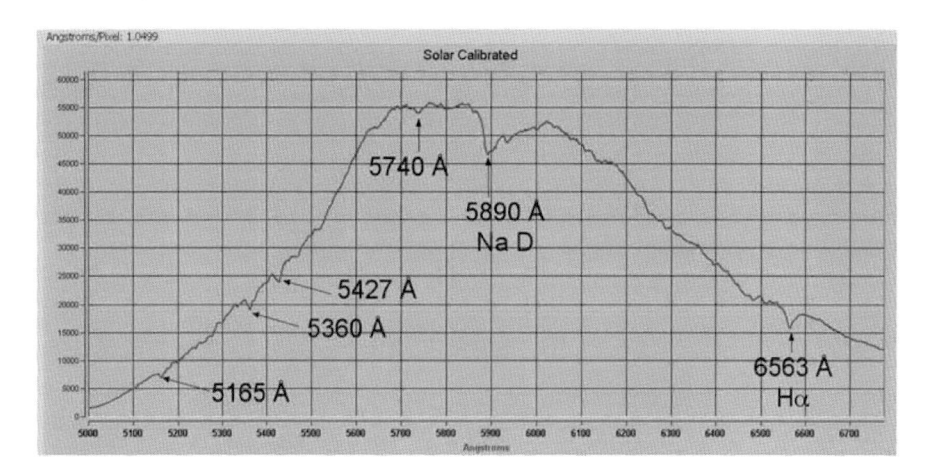

Fig. 3.15 Solar (G2V) line profile

Vega, Alpha Lyrae

HR 7001, HD 72167, SAO 67174; Apparent Magnitude V=0.03
RA 18 h 38 m 56.3 s, Dec +38 d 47′ 01.2″ (2000)

Perhaps the most popular star for doing spectroscopy, at least during the learning process, is the bright star Vega. Vega has an absolute magnitude of 0.58. Most of the hydrogen Balmer lines can be easily seen and identified making the star an ideal choice to learn on. Vega was the focal star in the movie "Contact" and is locate 25.04 light years from Earth (Fig. 3.16).

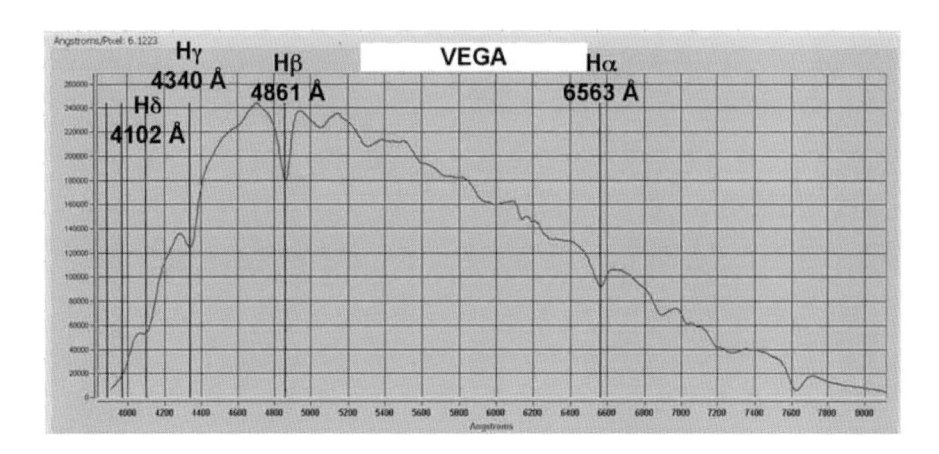

Fig. 3.16 Vega (alpha Lyrae – A0 V) line profile

Aldebaran, Alpha Tauri

HR 1457, HD 29139, SAO 94027; Apparent Magnitude V = 0.75–0.95
RA 04 h 35 m 55.2 s, Dec +16 d 30′ 33.5″ (2000)

Aldebaran is an easy star to find and is a giant red star located 68 light years from Earth. It has a 0.2 magnitude variation in the visual band (Fig. 3.17). Aldebaran has an absolute magnitude of −0.63.

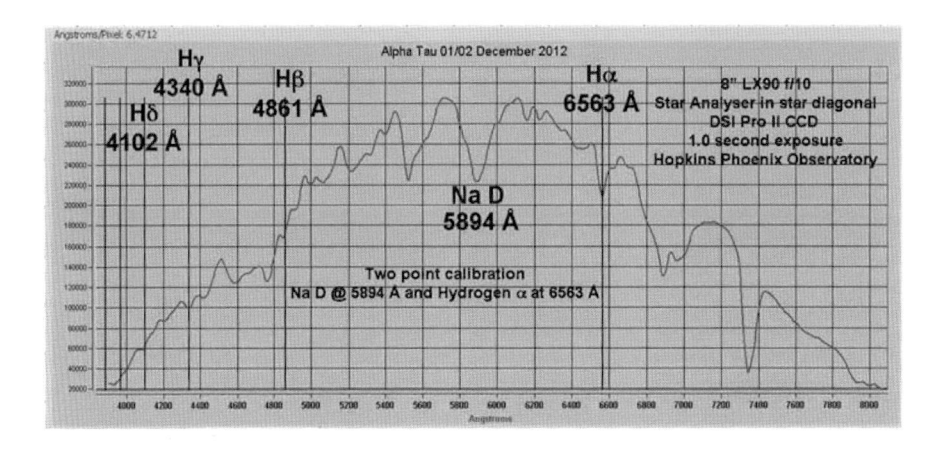

Fig. 3.17 Aldebaran (alpha Tauri – K5 III) line profile

Capella, Alpha Aurigae

HR 1708, HD 34029, SAO 40186; Apparent Magnitude V = 0.91
RA 05 h 16 m 41.4 s, Dec +45 d 59′ 52.8″ (2000)

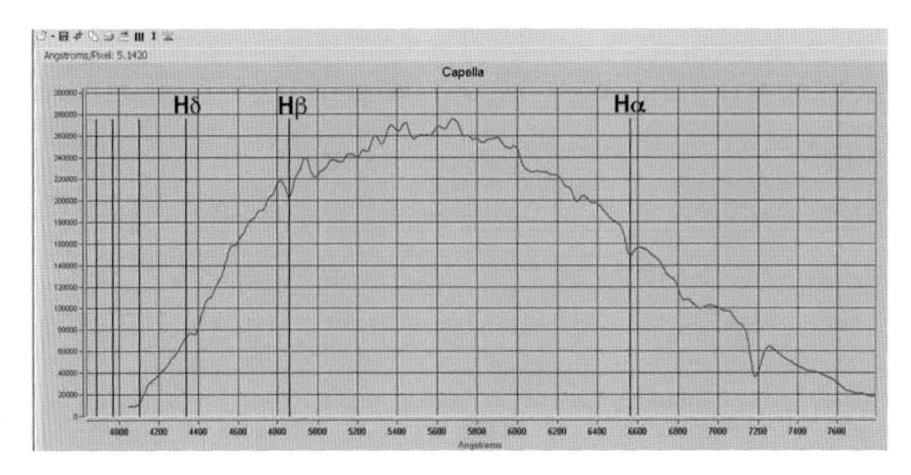

Fig. 3.18 Capella (alpha Aurigae – G8 III:+F) line profile

Capella is actually a four star system with two pairs of binary stars. Only one star can be seen, however. The brightest pair of binary stars has both G-type stars, each with a diameter of about 10 times the Sun. The absolute magnitudes of these two stars are 0.91 and 0.76. The second pair of stars is located about 10,000 AU from the first pair. The Capella star system is located about 42.2 light years from Earth. The hydrogen alpha and beta lines are prominent in the spectrum of Capella (Fig. 3.18).

Deneb, Alpha Cygni

HR 7924, HD 197345, SAO 49941; Apparent Magnitude V = 1.25
RA 20 h 41 m 25.9 s, Dec +45 d 16′ 49″ (2000)

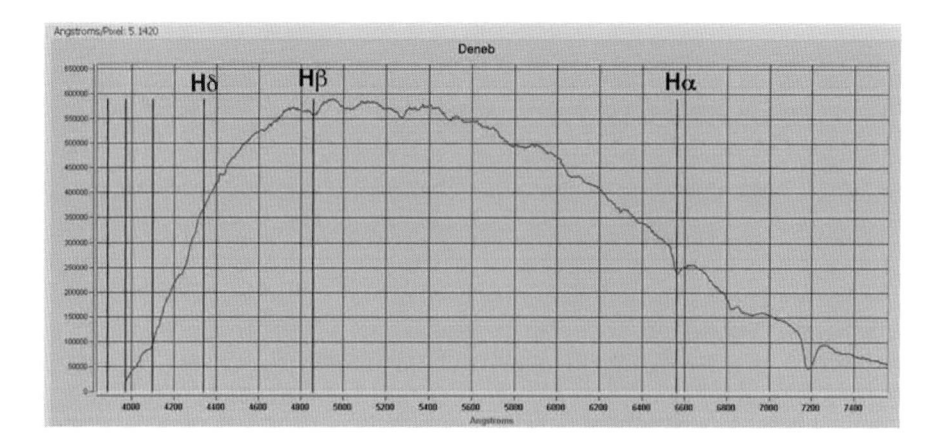

Fig. 3.19 Deneb (alpha Cygni – A2 Ia) line profile

Deneb is a blue-white supergiant and has a luminosity of nearly 200,000 times that of the Sun. Deneb has an absolute magnitude of −8.4 making one of the most luminous stars known. It is located about 2,600 light years from Earth. Deneb can be easily found and is at one of the vertices of the Summer Triangle (Fig. 3.19).

Betelgeuse, Alpha Orionis

HR 2061, HD 39801, SAO 113271; Apparent Magnitude V = 0.3–1.2
RA 05 h 55 m 10.3 s, Dec +07 d 24′ 25.4″ (2000)

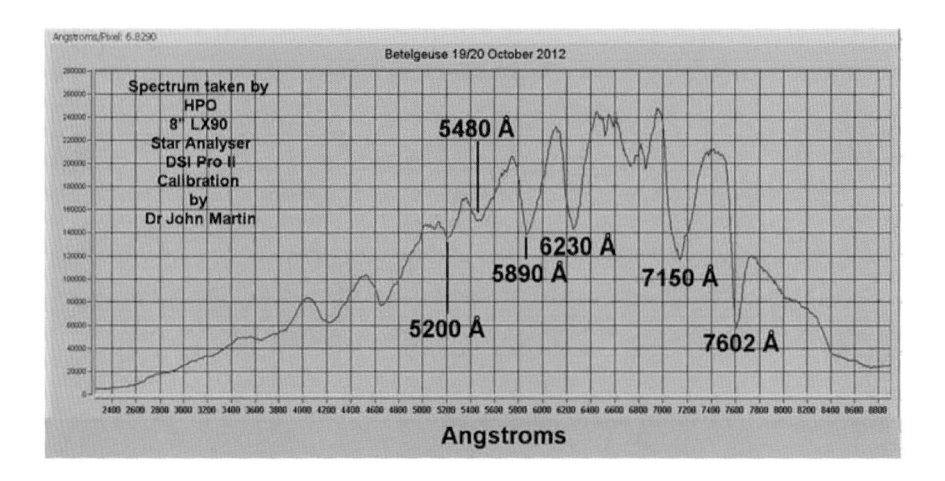

Fig. 3.20 Betelgeuse (alpha Orionis – M2 Iab) line profile

Betelgeuse is a red supergiant semi regular variable star with a short-term period of 100–300 days and a regular variation of approximately 5.7 years. The mass estimate for Betelgeuse is poorly known and estimates vary between 5 and 30 Suns. It has a luminosity of between 90,000 and 150,000 that of the Sun and an absolute magnitude of −6.02 making one of the most luminous stars known. It is located about 640 light years from Earth. Betelgeuse is one of the three stars that makes up the Winter Triangle and is the raise right hand of Orion. Betelgeuse is one of the suggested spectroscopic projects. Dr. John Martin (University of Illinois) was kind enough to provide a wavelength calibration for the following Betelgeuse spectrum line profile (Fig. 3.20).

Rigel, Beta Orionis

HR 1713, HD 34085, SAO 131907; Apparent Magnitude V = 0.05–0.18
RA 05 h 14 m 32.3 s, Dec –08 d 12′ 05.9″ (2000)

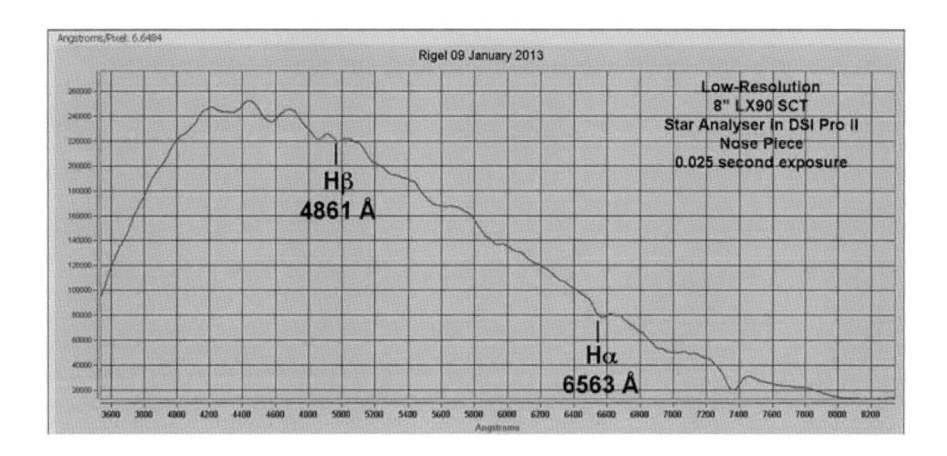

Fig. 3.21 Rigel (beta Orionis B8 Ia) line profile

Rigel is a triple star system with the brighter star, Rigel A, a blue-white supergiant star. The mass estimate for Rigel A is about 18 Suns and has a luminosity of 130,000 times that of the Sun. The star system's absolute magnitude is −7.84, mainly due to Rigel A, making one of the most luminous stars known. Rigel A is an irregular variable with periods varying from 1.2 to 74 days. The variation is due to pulsations of the star. The star system is located between 700 and 900 light years from Earth and is the left foot of Orion. Rigel B can be seen as a binary star with Rigel A, but because its magnitude is only V = 6.7, it can be a challenge to see next to Rigel A which is some 500 times brighter. Rigel B is a spectroscopic binary star system, but probably too much of a challenge for small telescopes. Figure 3.21 shows a wavelength-calibrated line profile of Beta Orionis (Rigel).

Bellatrix, Gamma Orionis

HR 1790, HD 35468, SAO 112740; Apparent Magnitude V = 1.59–1.64
RA 05 h 25 m 07.9 s, Dec +06 d 20′ 58.9″

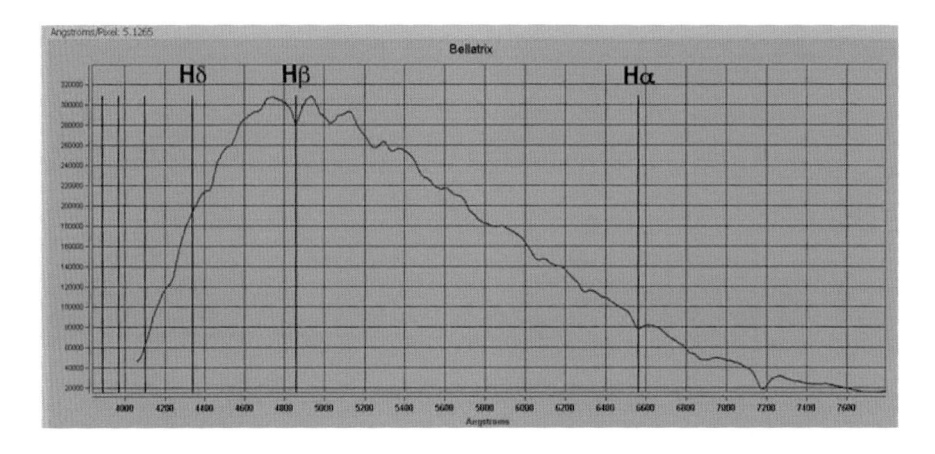

Fig. 3.22 Bellatrix (gamma Orionis – B2 III) line profile

Bellatrix is a blue-white giant star. The mass estimate for Bellatrix is about 8.4 Suns with a luminosity of 6,400 times that of the Sun. The star system's absolute magnitude is −4.2. The star system is located 250 light years from Earth and is the upper left hand of Orion (Fig. 3.22).

Sirius, Alpha Canis Majoris

HR 2491, HD 48915, SAO 151881; Apparent Magnitude V = −1.47
RA 06 h 45 m 08.9 s, Dec −16 d 42′ 58.0″

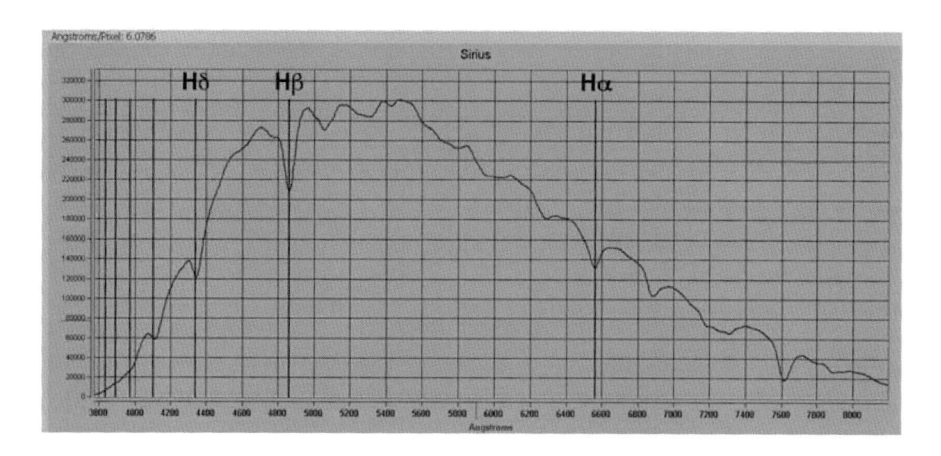

Fig. 3.23 Sirius (alpha CMa A – A1 V) line profile

Figure 3.23 shows a wavelength-calibrated line profile of Alpha Canis Major Sirius. Sirius, also known as the "dog Star" for being Orion's dog, is the brightest star in the sky as seen from Earth, aside for our star the Sun. This makes it an easy target from the northern hemisphere. Sirius is an optical binary with Sirius A the main star and a much fainter Sirius B at V = 11.18. Sirius A has a mass of 2.02 Suns and a luminosity of 25.4 Suns. Sirius is located 8.60 light years from Earth. Sirius is a good star to practice on and has very definable hydrogen Balmer lines. Because of its brightness, only sub-second exposures are required. In fact, good exposures that do not saturate and remain in the pixel ADU count linear range may be in just a few milliseconds depending on the size of the telescope. The hydrogen beta line is very pronounced with the hydrogen alpha not as much.

Star Analyser Conclusion

Te secret to successful spectroscopy with a Star Analyser is to experiment. It is very hard to hurt anything. Unlike the old film type cameras, digital cameras need no developing and printing of images. Results can be seen immediately. Taking a thousand images cost no more than one image, which aside from the camera cost. Even taking one image costs nothing. The more you do the better the results will be. Because there is no slit, what is to be imaged needs to be very small in order to create a spectrum. Stars work well, but even extended objects can produce a spectrum.

It is possible to get a spectrum of say M57, the Ring Nebula. Even spectral images of the planets, comets and the Moon are possible. What gets produced is what looks like a smeared color image of the object, but it is the spectrum of the object. The trick is to take the image and when processing the image just take a slice of the spectrum image and make a line profile of it. Remember most of the spectrum of solar system objects will be of light reflected from the Sun. This means most lines and features are that of the Sun, not the object. Careful study can reveal some of the objects own contribution to the spectrum, but that is no easy task. The Star Analyser is a great device to start learning about astronomical spectroscopy and at a very reasonable price.

A clear night is important, but even with clouds, as long as you can see the star you should be able to get a good spectrum image. As an experiment, some night when there are scattered clouds, pick a bright star and watch it's spectrum. See how clouds passing in front of it affect it. You may be surprised. A dark sky is preferred to a bright one, but still even in a light polluted suburban area with a full Moon close to the object, good spectra of bright stars can be taken. As with most astronomical work the best place for an object to be when observed is at or close to the zenith or at least on the meridian. The term Air Mass is used to define how much atmosphere light must transverse. By definition the air mass at the zenith is 1.00. The further from the zenith or closer to the horizon, the higher the air mass and more of the atmosphere the object's photons must transverse. Not all or even many object of interest will pass through the zenith. The location of the observatory will determine an object's highest elevation. The best place to observe an object is when it is on its meridian, a line from the horizon through the zenith. In spectroscopy there are two reasons for worrying about the air mass. The absorption of photons by the water molecules in the atmosphere can cause the addition of significant telluric lines (mainly due to water vapor) in the spectrum. This is true even in dry desert areas or tops of mountains. The second air mass concern is extinction. This is the attenuation of the photons by the atmosphere. The greatest extinction will occur at the higher air mass. To complicate things, at a given air mass, the greatest extinction occurs in the shorter wavelengths. Imaging the object when it is at its lowest air mass or closest to the zenith can minimize the difference in extinction from the blue to red end of the spectrum. As would be expected, a given exposure for an object will produce different ADU counts near the zenith versus near the horizon. Near the zenith there will be higher ADU counts and thus a better signal-to-noise ratio with observations made. Always try to plan to observe an object when it is close to the meridian.

Chapter 4

DIY Spectroscopy

Introduction

An interesting item has been presented on eBay several times over the last few years, a Do-It-Yourself (DIY) complete system spectrograph referred to as a spectrometer. Figure 4.1 shows the front view of the DIY spectroscope. It is actually a misnomer to call these units DIY. They are complete and while one can do fine tuning and wavelength window adjustments, there is little else to do other than to turn on connect to a computer and take spectra. These units are refurbished, but appear to be in like-new condition. They sell for less than $200, about the price of the Star Analyser. More information on them can be found on the Science-Surplus.

Fig. 4.1 DIY spectroscope, front view

com web site at http://www.science-surplus.com. The site indicated an aligned unit is $500, however, the three units tested were purchased for $200 each and were all aligned well. In addition, if a unit needs alignment, the procedure is not difficult. The web site also indicates discounts to educators. At under $200 many interested astronomers purchased these units. Most of the spectrometers come with an 1,800 l/mm grating. Some units come with a 900 l/mm grating. The ones with the 900 l/mm grating sold for $299, but otherwise are identical to the 1,800 l/mm grating spectrometer. Additional gratings can be purchased for $75.00 (Table 4.1).

Table 4.1 Available gratings

Grating	Wavelength range	Resolution
2,400 l/mm	3,100–4,500 Å	7.5 Å
1,800 l/mm	3,650–6,000 Å	10.0 Å
1,800 l/mm	5,000–7,000 Å	10.0 Å
900 l/mm	3,650–9,100 Å	20.0 Å
600 l/mm	3,650–1,100 Å	30.0 Å

The spectrometers are crossed Czemy-Turner design, built by B&W Tek (Model BTC-110S). As mentioned above these units are refurbished OEM spectrometers and formerly used as medical devices. They are completely self-contained. The spectrometer is fiber optic coupled and comes with a 0.5-m long 200 mm (0.2 mm) diameter step index fiber optic cable and SMA connectors. It appears to be a glass fiber as opposed to plastic. The transmission response has been tested and is good from 3,650 Å (365 nm) to 9,300 Å (930 nm). The spectrometers have a small optical bench built-in. The two mirrors on the optical bench can be adjusted to fine tune the optical path for optimum efficiency. The optical bench has an adjustable 1,800 l/mm grating (some units have a 900 l/mm grating). While this requires taking the unit apart and opening the optical bench, once exposed the grating can be loosened and rotated to change the wavelength coverage of the spectrum window. There is a 50 mm slit included as well that is located inside on the fiber optic port. The detector is a linear monochrome CCD chip (Sony ILX511, 1×2048 pixels, 14 mm by 28 mm). There is also a very nice Windows based computer interface and supporting software program called "Spectrum Studio" to control the spectrometer and process the data. The computer interface is via a serial cable. Newer computers may not have a serial port, but USB to serial adapters are available and are inexpensive. An included 5 VDC power supply (115 VAC) supplies power to the unit. The spectrum is small and measures 5.75″ (140 mm) × 3.75″ (95 mm) × 1.75″ (45 mm). The DIY spectrometer comes with the spectrometer, a serial computer cable, 19″ fiber optic cable with SMA connectors on both ends and a CD ROM with the software.

This unit is ideal for someone wishing to explore spectroscopy. Experiments are easy and can be done indoors and even in daylight. Pointing the fiber optic cable at various light sources, such as CFLs or different color lasers pointed at a piece of paper, can produce excellent spectrum line profiles. Pointing the fiber optic cable

DIY Spectrometer Specifications

outside during the day will produce a very nice spectrum line profile of the Sun. This allows practice with processing the line profile. The line profile can be wavelength calibrated and various elemental lines identified.

The wavelength sensitivity is best in the shorter wavelengths and goes well below 4,000 Å (Fig. 4.2). Even the longer wavelengths retain good sensitivity beyond 8,000 Å.

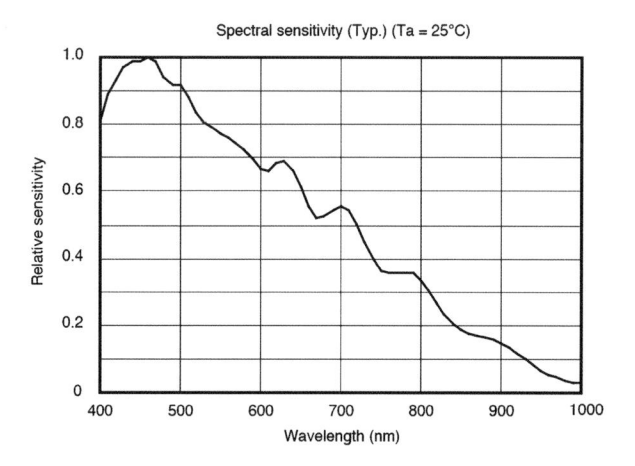

Fig. 4.2 Sony ILX511 CCD wavelength response

The included software Spectrum Studio is written in C# and requires at least Windows XP or Windows 7. It uses Windows NET framework 3.5, which is standard on Windows computers.

DIY Spectrometer Specifications

Exposures can be set from 50 to 65,535 ms. There is no spectrum displayed, only a line profile. Multiple line profiles can be summed and averaged. The number to be summed/averaged profiles can be selected from 1 to 1,000,000,000 (one billion). Dark frames are automatically subtracted. With the 1,800 l/mm grating a resolution of 10 Å or better can be obtained. This increased resolution is a big improvement over the Star Analyser and even the ALPY 600 at a small fraction of the cost. The line profile data can be saved and downloaded as a text file. By manipulating the text file and saving the resulting file with a .dat extension, the file can be opened in RSpec or VSpec and processed like a regular spectrum. The data can also be processed with the included program.

For anyone desiring such a spectrometer it is suggested that eBay be checked periodically. As noted above, these are refurbished units and may appear at random times over the next few years. Just when it looks like they are all sold out, another dozen or so are listed for sale.

DIY Spectrometer Equipment

One of the nice features of the DIY Spectrograph is no other equipment is needed other than a computer and a way to get light into the fiber optic cable. The biggest problems are for astronomical work. To couple the fiber optic cable to a telescope is no easy feat. Getting the starlight into the fiber and guiding to keep it there can be a challenge.

Fiber Optic Cables

There are a multitude of different fiber optic cables and connectors available. Most fiber optic cables are used for data communication. Because they allow light, including the visible spectrum, to be transmitted with little loss, they are also ideal for use with astronomical spectroscopy. For astronomical spectroscopy most any fiber optic cable will work, but the larger the diameter of the fiber, the better. Note if a spectrograph does not use a slit, the smaller fiber optic cables are better as they provide an effective small light beam like that produced with a slit. There are both high-performance glass fibers and more economical plastic fibers. The glass fibers are designed for long distance and high data rate communications. For spectroscopy the fiber optic cable need not be more than 2 m or 3 m in length. That means the plastic cables work fine. The glass fiber optic cables can cost in the hundreds of dollars while the plastic fiber optic cables can be purchased for under $10. There are very economical audio fiber optic cables that can be purchased from All Electronics (http://www.allelectronics.com/). Cables can be purchased in various lengths, $8.50 for a 20 foot cable, $5.50 for a 12 foot cable, $4.50 for a 6 foot cable and $2.50 for a 3 foot cable. These are cables with connectors on each end and are called TOSLINK (Toshiba Link) optical cables. For even the 12 foot TOSLINK cable no loss of spectrum intensity was seen when compared with a high-quality 18″ glass fiber optic cable that came with the DIY spectrometer. The TOSLINK cables are designed for minimum transmission loss at a central wavelength of around 6,500 Å. This makes them idea for the hydrogen alpha area as well as the helium I area, but the cables work well in the shorter wavelengths too.

What size cable should be used? The slit of a spectrograph is usually 50 mm wide or smaller. This means the cables can be a minimum of that size. For slitless spectrographs the smaller the slit the better, but this can complicate getting light into the fiber and keeping it there for an exposure. Most fiber optic cables are 100, 200 500 or 1,000 (1.00 mm) microns in diameter. The TOSLINK plastic cables mentioned above have a fiber that is 1.0 mm in diameter and seems to work fine.

What connectors should be used? The purpose of the connectors is twofold, to keep the fiber centered and to allow easy connection. The fiber optic cable that comes with the DIY spectrometer is a 19″ glass cable with SMA connectors. The SMA connectors are nice and provide solid and aligned connections. They are a

DIY Spectrometer Equipment

standard for RF microwave use and have been used without the brass pin in the center for fiber optic cables. The connector used is not critical and most any connector can be used or even no connector as long as the fiber optic cable can remain aligned. The TOSLINK plastic fiber optic cables come with a plastic connector and brass center tube encasing the fiber. The plastic connector on the TOSLINK cable can be removed and a RF type SMA connector drilled out to allow insertion of the brass center with plastic fiber into it the connector in place of the center pin. It would be nice if mating TOSLINK chassis connectors could be used, but trying to find mating TOSLINK connectors is futile. The reason for the problem is that all the TOSLINK chassis connectors that are readily available are designed to work as a system. The mating TOSLINK chassis connectors contain circuitry to convert between audio and optical signals. It is possible to purchase these chassis connector devices and just use the connector, but it is just as easy and much cheaper to modify the TOSLINK cable's connector (Fig. 4.3).

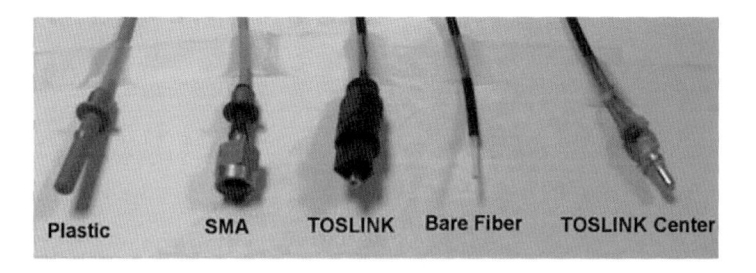

Fig. 4.3 Fiber optic cables and connectors

The fiber optic cable that comes with the DIY Spectrograph uses standard SMA optical connectors. To experiment with coupling the TOSLINK fiber optic cable to a telescope, the cable was used without the connector on the telescope end and inserted in a hole drilled in the star diagonal mirror. Drilling the hole in the glass mirror takes practice. Very small diamond tipped drills must be used. The nice thing is the diamond drills are not expensive (~$5.00 for 5) and can be obtained from Arrowhead Lapidary & Supply (http://www.arrowheadlapidarysupply.com).

It is suggested that some practice holes be drilled in glass slides or other pieces of glass before doing the star diagonal mirror. A small drill press should be used. For the experiments at the Hopkins Phoenix Observatory a small Unimat was configured as a drill press. The Unimat is a multiple tool device that can be configured as a drill press, lathe, grinder or mill. They are no longer available new, but some are available used, but at a high price, upward to $1,000. Larger drill presses will be difficult to use, as they normally cannot hold the small drills. If something like the Unimat is not available, there are multiple small drill presses available ranging from under $100 to several hundred dollars. An alternate method is to use a Dremel or similar tool mounted in a drill press configuration. To drill the hole, medium speed should be used with light but continuous pressure. The hole and drill must be lubricated. Water, WD-40 or isopropyl alcohol can be used. The alcohol seemed to work best and left no residue. Since there is very little heat and only small amounts are needed, fire is not a concern. It would be best not to smoke or have open flames close by, however. To drill the mirror, first remove it from the star diagonal and then mount it on the backside of the star diagonal with the back of the mirror exposed. Put the star diagonal in a vise with the telescope port down. This will then provide the proper alignment for the hole. Remember drilling at the center on the back will not place the opening on the mirror side in the center. The hole must be offset to come close to the center. Check the mirror thickness and make an estimation of how far offset must be. To prevent break through chips, bee's wax or another thin piece of glass on the mirror side will help. If a 1.0 mm fiber is used the hole can be 1.0 mm. Note that when breaking through the mirror side, it is very easy to cause a large chip to flack off. If this happens another mirror must be drilled or a small first surface mirror with a 1.0 mm hole in it glued over the main mirror hole. This cannot be hurried. Another suggestion is if the mirror flacks, use a small piece of aluminum foil glued to the mirror with a 1.0 mm hole in it. While the optical quality of the aluminum foil is poor it should be sufficient to allow the star's image to be seen and moved to the hole (Fig. 4.4).

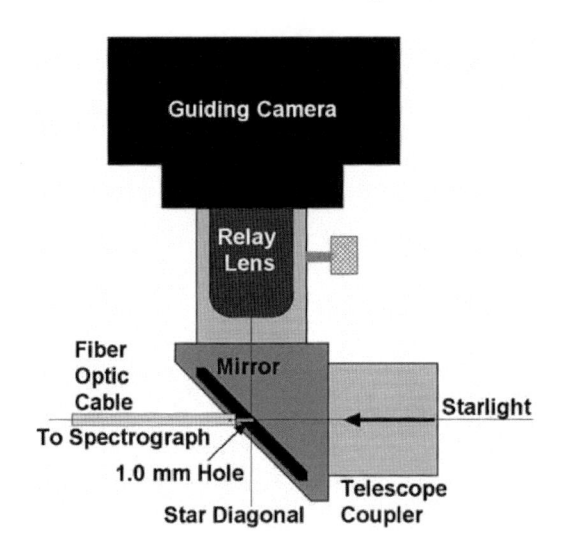

Fig. 4.4 Star diagonal detail

The next big challenge is getting a star's light into the fiber.

In order to be able to focus on the hole in the mirror, a relay lens is needed. If an eyepiece is used for guiding, the relay lens is inserted into the eyepiece barrel. The larger lens should be toward the mirror. Focusing can be accomplished by adjusting the position of the relay lens in the barrel and by moving the whole eyepiece assembly in and out. A 1.0" aluminum rod was used for the lens mount, but any material such as plastic or even wood could be used. This is best made using a small lathe, but a drill press could also be used (Figs. 4.5 and 4.6).

Using and Adjusting the DIY Spectrometer 107

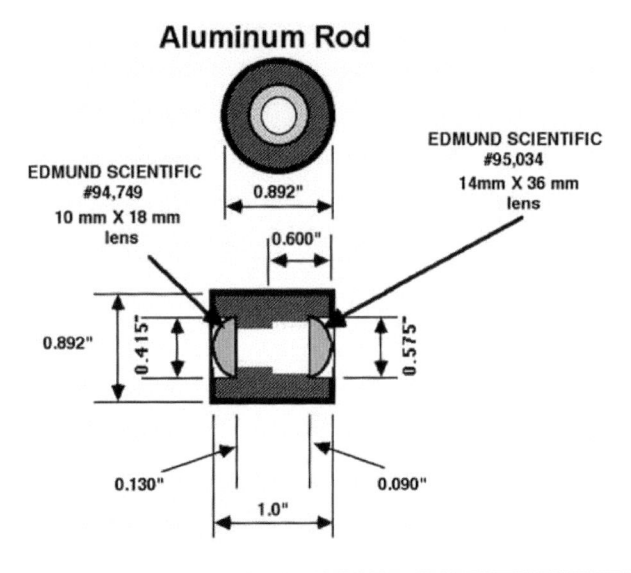

Fig. 4.5 Relay lens detail

Fig. 4.6 Star diagonal and relay lens

Using and Adjusting the DIY Spectrometer

Introduction

The DIY spectrometer is pretty much plug and play. As mentioned above the biggest problem for astronomical use is to find a good way to have the astronomical object's light injected into the fiber optic. Again, this is harder than it may seem. For calibration lamps or solar spectra, just pointing the fiber at the source works well. For most astronomical objects a telescope will be needed and getting the focused image into the fiber optic is a large challenge. It is suggested the user experiment with the above methods as this can be done with a minimum of expense. For those observers who do not like to make things, Shelyak sells a telescope fiber optic coupler that is normally used with their eShel spectrograph. That telescope coupler should work fine, but at $3,700, the cost may be prohibitive. Shelyak also sells a coupler for the ALPY 600 for around $100. This is not a telescope coupler and only couplers a fiber optic cable to the ALPY. This is rather expensive considering what is needed.

To use the spectrograph, connect the DIY spectrometer to a Windows based computer using the included serial cable. Make sure the Software that comes with the unit is loaded and opened. Connect the 5 VDC power supply that comes with the unit to an AC source and plug it into the DIY spectrometer in the DC 5 V port. The LED on the unit should illuminate yellow/green and then turn red. There is a fan inside that has been disconnected. Since it will help the cooling of the CCD chip and increase the sensitivity it suggested it be reconnected.

Optical Bench

The DIY Spectrometer uses a very nice small optical bench. To access it there are 12 small screws that must be removed for making adjustments in the optical bench. The screws are very small so be sure to put them in a secure location. They can be lost very easily. To remove the blue cover, remove the four small Philip's head screws on the sides of the spectrometer. Note, if the fiber optic cable is attached it must be removed to get the cover off.

Note: The electronics inside the unit are sensitive to electrostatic discharges. Standard grounding procedures should be exercised.

Once the cover has been removed the optical bench, a 2.5″ square black box, can be accessed. Removing the cover will allow access to the inside of the optical bench. Four small hex screws secure the cover. A 0.050″ hex wrench is used to remove these screws.

Note: The cover is asymmetrical and can only be installed in one position.

With the cover removed adjustments to the mirrors and grating can be made. Perhaps the most frequent adjustment is for changing the spectrum window position. To do this the diffraction grating can be rotated to cover different sections of the visible spectrum. The mirrors can also be adjusted to maximize performance. Information for doing these adjustments is detailed below. The light path for the spectrum can be seen in the following Fig. 4.7. The light passes from the fiber optic cable through the SMA chassis connector and into the 50 mm slit. From there the light beam goes to a collimating mirror that collimates the light and sends it to the diffraction grating. The spectrum produced by the diffraction grating is then sent to the focusing mirror. The focusing mirror sends the focused spectrum to the linear CCD.

Using and Adjusting the DIY Spectrometer

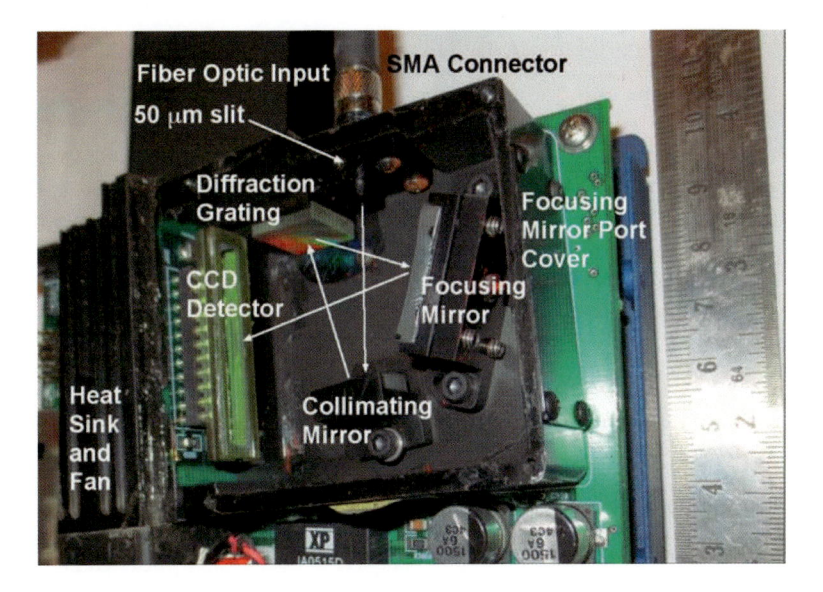

Fig. 4.7 DIY spectrograph optical bench

Optical Bench Alignment

There are two mirrors on the optical bench. Their purpose is to place an image of the entrance slit on the plane defined by the CCD detector. When the mirrors are not properly aligned, the image plane is curved. If this is the case, spectral lines at the low-numbered pixel positions will be sharp with the spectral lines at the other end broad. The curvature of the focal plane is dependent on the position of the two mirrors. This is a very sensitive relationship. The spectrometers tested at the Hopkins Phoenix Observatory were all well aligned, but could probably have been adjusted for peak performance. If the spectrometer is not aligned well or if the user wishes to fine-tune the optical bench for a specific wavelength or for better overall spectrum response, the following procedures can be done. The order of adjustments is important. First adjust the collimating mirror, then the diffraction grating and finally the focusing mirror. A final spectrum window adjustment and fine-tuning of the collimating mirror can then be done. The first adjustments are done in a lighted environment and with power to the spectrometer off. The last adjustments require the spectrometer to have power on and be producing a line profile. This last part must be done in a darkened environment. To begin, the inside of the optical bench must be accessed as described above. The orange fiber optic cable supplied must be reattached to the unit. Hold a small piece of white paper behind the slit, on the inside of the optical bench. Shine a light from a red or green laser into the fiber optic cable. The light should be seen on the paper as a dot. If no light is seen, check the fiber optic cable and slit for any obstructions or other problems. Once it the light is seen on the paper, remove the paper.

Collimating Mirror Adjustment

There will be two adjustments of the collimating mirror. The following is the preliminary adjustment. To adjust the collimating mirror loosen the one screw holding it using a 3/32″ hex wrench. This allows the collimating mirror to both rotate and moved in and out. For this adjustment, only the rotation is done. Do not move the mirror in or out. With the laser light hitting the collimating mirror adjust its rotational position so the light hits the diffraction grating on the side closest to the slit, about 15 % from the edge. Tighten the screw holding the collimating mirror.

Diffraction Grating Adjustment

There are two different sized screws holding the diffraction grating in place. There is a small screw toward the inside. This uses a 0.050″ hex wrench. A larger screw is toward the wall of the optical bench. This screw uses a 3/32″ hex wrench. The key here is the laser light should be showing in the focusing mirror. If it is not, slightly loosen these two screws. Do not remove these screws. Adjust the grating by grasping its edges of the and slightly rotating it slightly until the light is seen in the focusing mirror. Now retighten the two screws.

Focusing Mirror Adjustment

There are several adjustments for the focusing mirror. The mirror can be rotated horizontally as well as tipped and tilted in the vertical plane. To adjust the focusing mirror the port cover on the side of the optical bench housing behind the focusing mirror must be removed. The 0.050″ hex wrench is used to remove the four screws holding the cover.

Note: The four screws are the same size as the screws that hold the optical bench cover on, but considerably shorter.

Removing the port cover will allow access to the adjusting screws on the back of the focusing mirror. Mirror rotation is accomplished by slightly loosening (do not remove) two 0.050″ hex screws at the ends of the focusing mirror. The object is to put the laser dot on the CCD detector's sensitive pixel array area. This area is a thin gray line running horizontally across the detector. While the rotation adjustment of the mirror is straight forward, the tip and tilt adjustment requires the loosening of the opposite screw before tightening an opposite screw. The mirror is suspended with a bolt in the middle and a spring to keep the mirror under tension. This is why the opposite screw must be loosened the same amount as the screw is tightened. Make small adjustments. The object is to put the laser dot on the pixel area midway from top to bottom of the area and approximately where in the spectrum the wavelength of the laser should be. In other words, if a red laser is used the red dot should be centered vertically and about 80–90 % of the way to the right (toward the slit end of the optical bench). A green laser should have the dot centered vertically and about midway from right and left. Tighten the two 0.050" hex screws holding the focusing mirror. Do not touch the four screws on the back of the mirror. Put the focusing mirror port cover back on the optical bench housing using the four shorter screws and the 0.050" hex wrench.

Spectrum Window Adjustment

This adjustment should be done after the alignment as indicated above is satisfactory. The purpose of this adjustment is to move the spectrum window across the CCD detector to view different areas of the visible spectrum in detail. This adjustment is most important for the 1,800 and 2,400 l/mm gratings. This is because the detector will not be able to image the complete visible spectrum with these gratings. Lower resolution gratings will be able to image the complete visible spectrum so the only adjustment for those would be to center the spectrum in the detector. This adjustment requires a multiple line spectrum light, such as a CFL being shown into the fiber optic cable. To change the wavelength window the diffraction grating is rotated slightly to change the angle of the grating relative to the optical path. Changing the position of the diffraction grating will allow selection of the desired part of the visible spectrum.

Note: Never touch the surface of the grating. It can be damaged very easily. Always grasp the grating by the edges. Never try cleaning the surface. If there is dust on it, use a gentle stream of clean dry air to blow the dust off.

As noted above for the diffraction grating adjustment there are two screws securing the grating. There is one 3/32" hex screw and one 0/050" hex screw. Loosen both slightly. Before proceeding, connect the spectrometer to a computer and apply the 5 VDC. Load the Spectrum Studio software, detect the spectrometer and select the "Continuous Scan Mode" to create a line profile. Keep the default 50 ms Integration Time and Averages of 1. See section "Spectrum Studio Software" for more information on using the software. Because the CCD is very sensitive and since the optical bench cover is still off, the following procedure must be done in a darkened environment. Shine the laser light into the fiber optic. A large spectral line should be seen in the line profile. Adjust the grating until the line is about where it should be, depending on the wavelength of the laser. Once satisfied, tighten the two screws on the grating.

Final Optimization

Now the collimating lens can be fine-tuned. This adjustment requires a multiple line spectrum light, such as a CFL being shown into the fiber optic cable. The computer program must still be running and producing a line profile. The collimating mirror can now be fine–tuned to produce the best line profile. Moving the collimating mirror in or out makes this adjustment. This is an extremely fine adjustment and a considerable challenge to get it correct. Loosen the collimating mirror hex screw again. Be very careful not to rotate the mirror. While monitoring the computer screen's line profile move the collimating mirror in and out to maximize the line profile. The spectral lines should be sharp at both ends of the spectrum. Once satisfied, tighten the mirror's hex screw. Once the adjustments have been made the power should be removed from the unit and the optical bench cover and spectrometer covers reinstalled.

Taking a Spectrum

Once the spectrometer is adjusted satisfactorily, it is time to take a spectrum image and produce a line profile. It is suggested that to start experimenting the user should keep things simple. By just holding the fiber optic cable in your hand and aiming it at various light sources a lot of experimenting can be done. This experimentation allows familiarization with the spectrograph and software. It also allows various spectra to be analyzed and wavelengths determined. The following Fig. 4.8 shows a mercury spectrum line profile from a florescent light. The fiber optic cable can also be pointed toward a window during the day and a nice solar spectrum line profile produced.

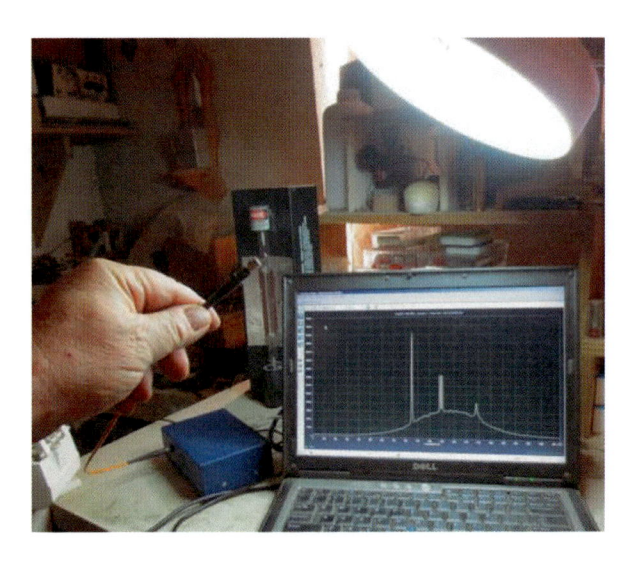

Fig. 4.8 Mercury spectrum line profile

Taking a Spectrum

To take a stellar spectrum the fiber optic cable must be coupled to a telescope. An eyepiece with relay lens as described earlier can be used to find the star and guide, but an imaging camera may prove easier to guide with (Figs. 4.9 and 4.10).

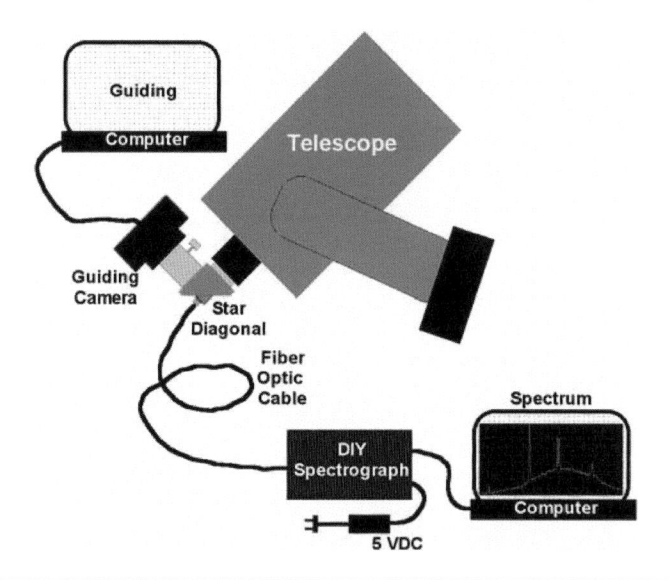

Fig. 4.9 DIY spectrograph for astronomy

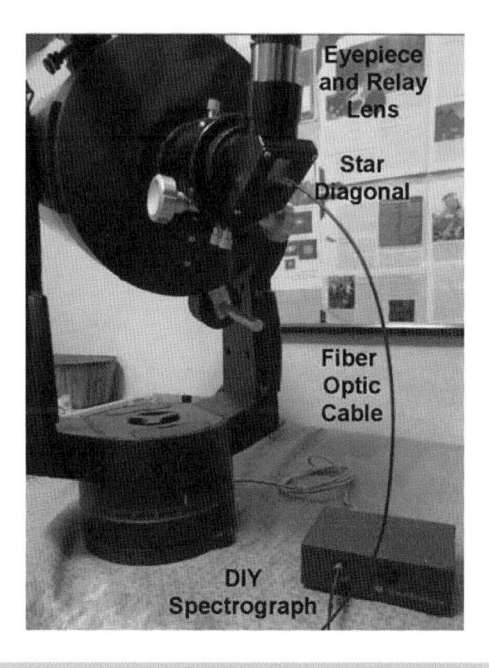

Fig. 4.10 DIY fiber optic interface

Spectrum Studio Software

The software that comes with the DIY spectrometer works on Windows and requires .NET framework 3.5 to be installed. It works fine with Windows XP and newer operating systems. This software is rather deceptive in that it may seem simple, but has some very advanced and powerful features. Perhaps the first thing that will be noticed is that there will be no spectrum image displayed. The software handles all the preprocessing of the spectrum image, such as dark frame subtraction, and creates a line profile from that information. The following information describes the features and options of the software in detail. For the software to work the spectrometer must be connected to the computer, via the serial port, and powered up by connecting the 5 VDC power supply. There is an LED on the spectrometer that will turn yellow/green when power is first applied and then red once everything is working properly.

With no power or if the spectrometer is not connected to the computer, the software will display a program window that is essentially blank and at the top of the screen the following will be displayed, "Spectrum Studio – No Spectrometers Detected." Once the DIY spectrometer is connected and working properly a spectrometer can be selected by clicking on the menu item at the top of the screen called "Spectrometer." The program should then automatically detect the spectrometer. The screen will then change to that seen in Fig. 4.11.

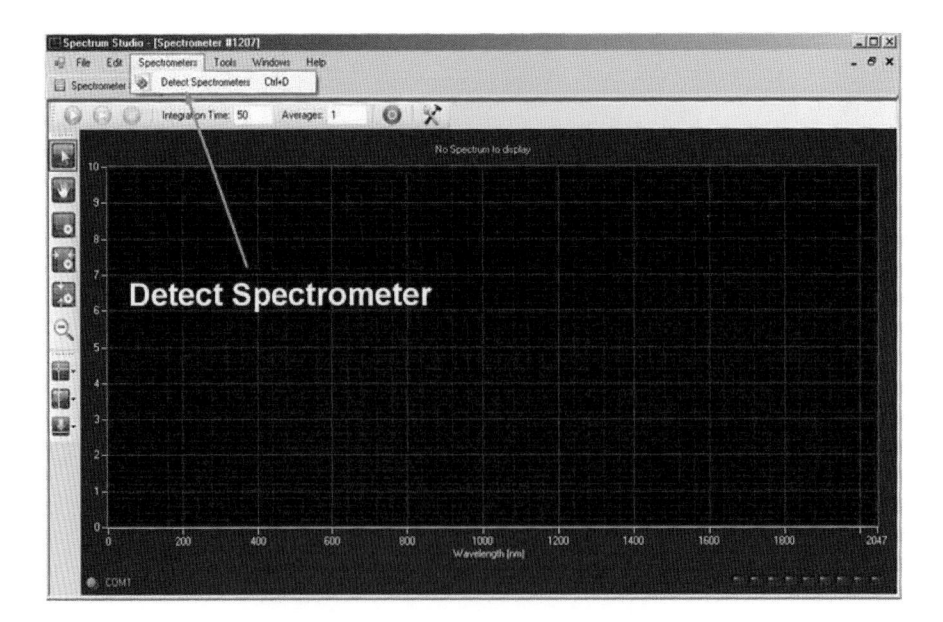

Fig. 4.11 Spectrum studio window

Spectrum Studio Software 115

Figure 4.12 shows an image of the window with various parts and icons defined.

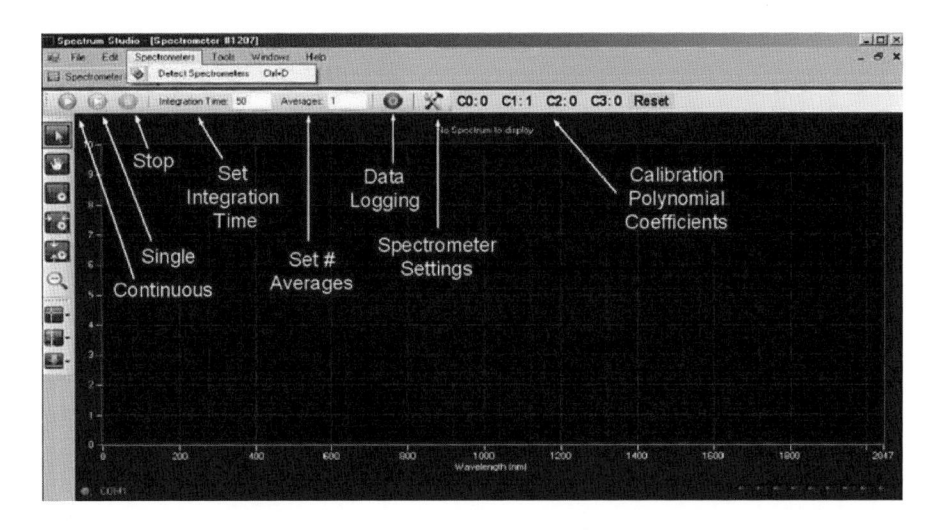

Fig. 4.12 Spectrum studio window tabs and icons

To see a line profile, click on the Single or Continuous buttons at the upper left. If the Continuous mode is selected and changes, such as the exposure length (Integration Time) or Averages or most other settings, first click on the Stop icon, the right most icon of the three at the upper left of the window. Even trying to quit the program can be a problem if the Stop has not been selected. The Integration Time can be set from 50 ms, default, to 65,535 ms. Multiple profiles can be averaged. The number to be averaged can be set from 1 default, to one billion.

The following Fig. 4.13 shows the Icons on the left of the window and their meanings.

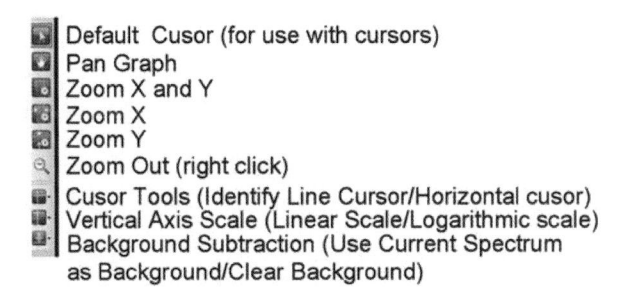

Default Cusor (for use with cursors)
Pan Graph
Zoom X and Y
Zoom X
Zoom Y
Zoom Out (right click)
Cusor Tools (Identify Line Cursor/Horizontal cusor)
Vertical Axis Scale (Linear Scale/Logarithmic scale)
Background Subtraction (Use Current Spectrum as Background/Clear Background)

Fig. 4.13 Spectrum studio window icons

Pan Graph
Selecting the Pan Graph allows using the left mouse/track pad button and cursor to drag the line profile window around. Right clicking the mouse/track pad button will return the screen back to normal.

Zoom X and Y
Selecting the Zoom X and Y allows the user to draw a box around the region to be zoomed using the left mouse/track pad button. Right clicking the mouse/track pad button will return the screen back to normal.

Zoom X
Selecting Zoom X allows the user to use the left mouse/track pad button to draw a vertical box around an area. The area in the box will then be expanded to fit the screen. Right clicking the mouse/track pad button will return the screen back to normal.

Zoom Y
Selecting Zoom Y allows the user to use the left mouse/track pad button to draw a horizontal box around an area of interest. The area in the box will then be expanded to fit the screen. Right clicking the mouse/track pad button will return the screen back to normal.

Zoom Out
Selecting Zoom Out has the same effect as right clicking the mouse/track pad button and will return the screen back to normal.

Cursor Tools
There are two options for the Cursor Tools, Identify Line Cursor and Horizontal Cursor. Note that these selections can be made together or separately. They do not interact with each other.

Identify Line Cursor
Selecting the Identify Line Cursor creates a yellow vertical line that can be dragged to the right or left. With a calibrated Line Profile the position of the yellow line is displayed in nanometers and the intensity in ADU counts. For example: (1,090, 3,600), where 1,090 is the X-axis wavelength at that position (1,090 nm) and 3,600 is the Y-axis intensity ADU count.

Horizontal Cursor
Selecting the Horizontal Cursor creates a blue horizontal line that can be dragged to the up and down. There is no readout, but the blue line intersects the vertical Y-Axis and its position can be easily seen.

Vertical Axis Scale
Selecting Vertical Axis Scale shows a popup menu with two selections, Linear Scale and Logarithmic Scale. This allows the Y-axis, vertical axis, to be calibrated linearly (e.g., 100, 200, 300, 400, 500 …) or logarithmically (e.g., 10, 100, 1,000, 10,000 …). The logarithmic allows low values to be seen along with very large values of intensity. Linear Scale is the default selection.

Background Subtraction
There are two sub-selections available when selecting this icon, Use Current Spectrum as Background and Clear Background. This option should not be normally used.

Spectrum Studio Software

Wavelength Calibrating the Line Profile

To wavelength-calibrate the profile the Tools icon is selected. The key to wavelength-calibrating the line profile is to determine the coefficients of an equation that will associate the proper wavelength for each pixel position. Remember, there is only one row of pixels and 2,047 columns or in other words just 2,047 pixel X-axis positions. The program uses a third-order polynomial for the wavelength-calibration. Four coefficients (C0 is the constant offset term, C1 is the linear term, C2 is the second-order term and C3 is the third-order term) for the calibration polynomial are displayed. By typing new numbers in the box that held the original value, the coefficient values can be changed. A program like Excel, as explained earlier can be used to determine these coefficients. Sometimes these coefficients are called **a0, a1, a2, a3**.

The default coefficients are: **C0 = 0, C1 = 1, C2 = 0, C3 = 0**

Once the calibration coefficients have been entered the X-Axis will be wavelength-calibrated in nanometers.

After a line profile has been created the top of the window changes from "No Spectrum To Display" to displaying the Integration Time, # of Averages and the computer's date and time when the spectrum was taken and line profile created (Fig. 4.14), "Integration Time: 50 ms; Averages: 1; Taken: Jul 17, 2013 at 10:3:26 PM."

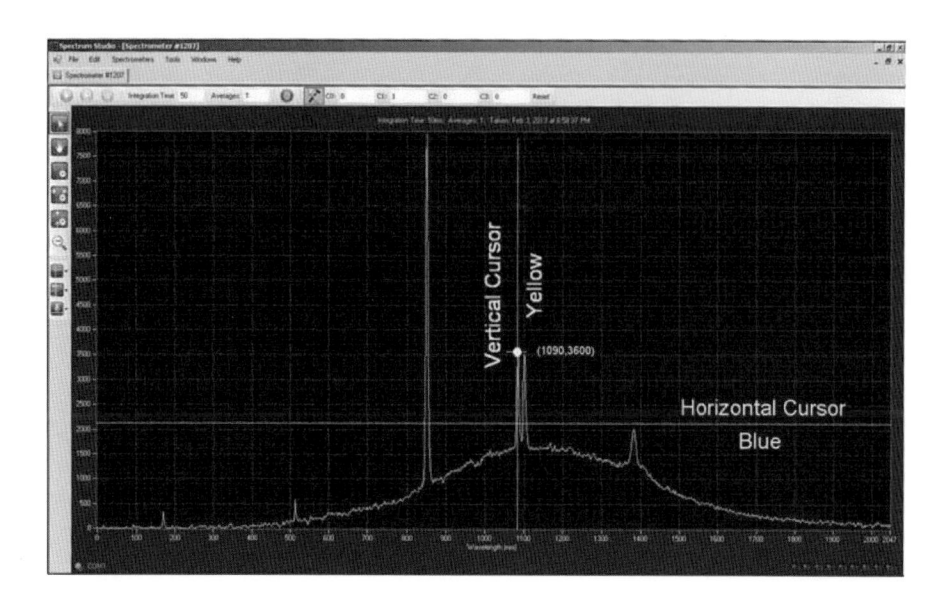

Fig. 4.14 Spectrum line profile

118   4   DIY Spectroscopy

Selecting the Tools icon and then the Option tab allows setting of Data Logging options and Spectral Line Identifiers.

Data Logging

Spectrum Studio can run in a data-logging mode where it saves every spectrum to disk as well as displaying them on screen. This is not something that would normally be done when doing astronomical spectroscopy. It could be utilized if there are multiple spectra to be taken, however. The data can either be written to a single file (one row per spectrum) or each spectrum in a separate file. From the Tools tab and Options window, the Data Logging tab can be selected. This allows several options for specifying the file naming and path to be used for data logging. Other options are also provided. Close that window. To enable data logging, click the "Record" button, the round blue button on the toolbar just to the right of the Averages box (as seen in Fig. 4.12). When Data Logging is active, there will be a small red dot is displayed in the middle of the blue button. Each completed spectrum will then be written to the computer's hard drive. To disable recording, push the "Record" button again so that the red dot turns gray. If the data file path does not exist, or the specified data file already exists, logging will be disabled (Fig. 4.15).

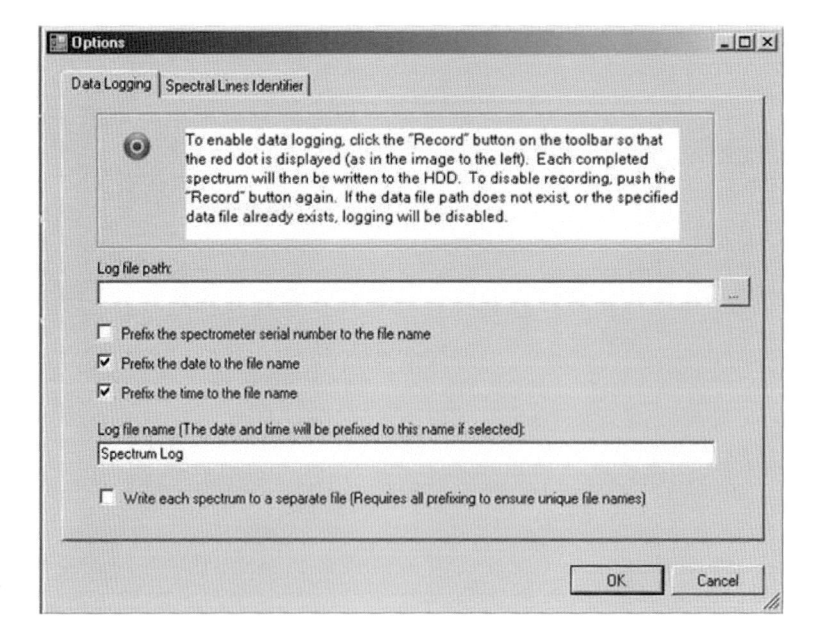

Fig. 4.15 Options – data logging

Spectral Lines Identifier

Spectrum Studio includes a line identification utility. Using wavelengths compiled and critically evaluated by scientists at the National Institute of Standards and Technology (NIST) in the NIST Atomic Spectra Database, this utility lets you identify spectral lines as described in the cursors section. To use this utility and because there are so many possible spectral lines, Spectrum Studio requires selection of which atoms or molecules that should be considered with the Identify Lines cursor. From the Tool tab at the top left of the screen select Options. Select the upper right tab, "Spectral Line Identifier." Next, click the check boxes next to the elements desired for consideration with the identification process (Fig. 4.16).

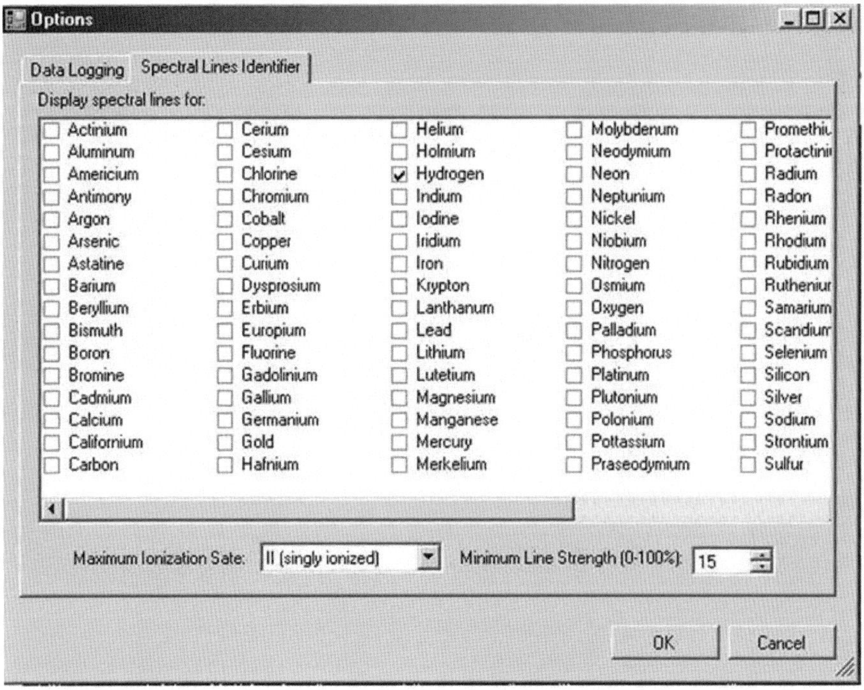

Fig. 4.16 Options – spectral line identification window

Once the elements have been selected close the Spectral Line Identification Window. From the set of icons on the lower left of the window, select the third from the bottom, "Cursor Tools." Two more selections will be shown. Click on the "Identify Line Cursor" option. The yellow vertical Identify Line Cursor will be displayed and the area to the right will show a white column. The element lines found by the yellow cursor will be displayed. In the following Table 4.2, the hydrogen alpha doublet is identified.

Table 4.2 Element identification

	Wavelength	Molecule	Strength
☑	656.272	Hydrogen I	67
☐	656.28	Hydrogen I	100

The wavelength and intensity are shown on the screen next to the cursor in yellow. If one of the check boxes is checked in the display to the right, that element will be spelled out on the window by the cursor in blue (Fig. 4.17).

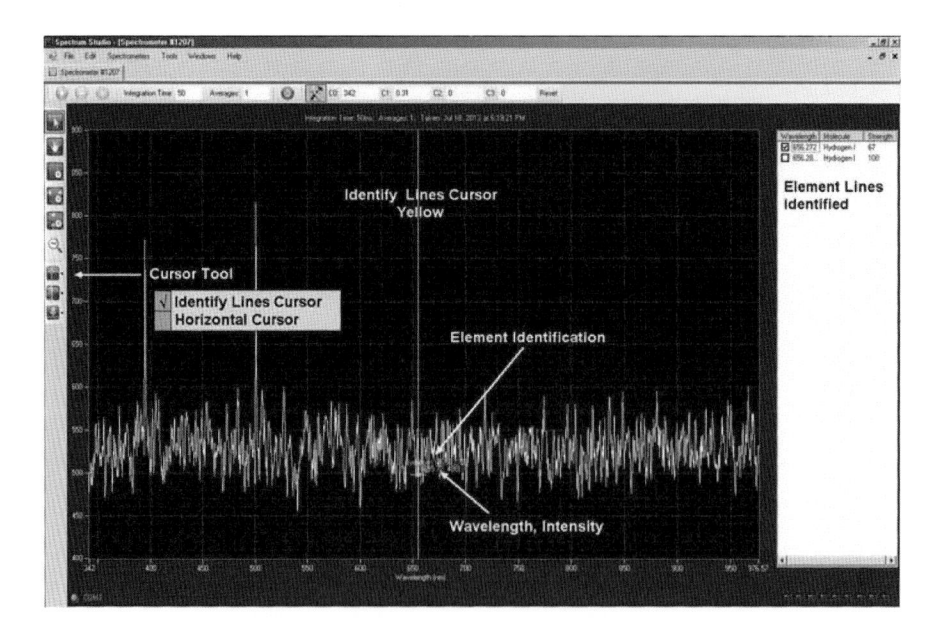

Fig. 4.17 Spectral line identification

Spectral Line Data Entry

The spectral lines data is stored in the "Lines" directory in path where Spectrum Studio was installed. Molecular lines or other spectral features can be added to the database. Adding files can be done by simply adding or modifying the files in this directory. Each file must be ASCII text with comma-delimited values (Notepad works fine for creating files if they are typed by hand). The name of the file must be "element-name.lines.csv" where element-name is the name you would like to appear in spectrum studio (e.g. hydrogen is "hydrogen.lines.csv"). Each line in the file represents a spectral line for the molecule. The first field is the wavelength in nanometers. The second field is the ionization state (0 is not ionized, 1 is ionized to the 1+ state, etc.). The third field is the strength of the line relative to other lines in the spectrum. This field must be a number between 0 and 100. For example, the first few lines of the Argon.lines.csv file are

212.516,2,40
213.387,2,60
213.859,2,40
214.873,2,40
216.619,2,60
216.826,2,40
217.023,2,80
217.722,2,100

The first line indicates that Argon 2+ has a spectral line at 212.516 nm with a relative strength of 40, and so forth. The spectral lines files are read into memory each time Spectrum Studio starts, so you will need to restart the program if you add or modify the files.

Interface Programming

Users can write their own interface. The following is from the Spectrum Studio software help file.

The serial interface is a standard RS-232 with the following parameters.

Baud Rate: 9,600 bps
Data Bits: 8 bits
Parity: None
Stop Bits: 1
Flow: Control: None

Commands:

Command	Description
A{Data}	Set the number of spectra to average before returning the averaged spectrum. Default value for {Data} is 1
	For example, to set the number of spectra to average to 10, use the command: **A10**
	Note: This feature is not used in Spectrum Studio because the spectrometer cannot communicate until the averaging process is finished. The average is performed in the computer instead
?A	Query the current average number
I{Data}	Sets the integration time to the value specified {Data} (in milliseconds). The valid range is 50–65,000 ms.
	For example, to set the integration time to 75 ms, send the command: **I75**
?I	Query the integration time
K{Data}	Sets the baud rate to the value specified by {Data}. Valid values of {Data} are:
	7 = 600 baud
	6 = 1,200 baud
	5 = 2,400 baud
	4 = 4,800 baud
	3 = 9,600 baud
	2 = 19,200 baud
	1 = 38,400 baud
	0 = 115,200 baud
	The default baud rate is 9,600 (3)
	For example, to set the rate to 115,200 baud, send the command: **K0**
?K	Query the current baud rate
Q	Reset the spectrometer. This resets all parameters to their default values
S	Initiate a scan and return the scan values. Scan values are returned as either ASCII text numbers or binary data, depending on the mode (see below)
a	Sets the spectrometer to communicate in ASCII mode. This mode is easiest to interpret using a serial terminal program, but it requires that more bytes are transmitted. This can make scan reporting quite slow
b	Sets the spectrometer to communicate in binary mode. In this mode scan data is sent in the numerical value of the bytes rather than encoded in ASCII numbers. The scan data is also compressed using the algorithm described below to decrease communication time
?a	Query whether the spectrometer is in ASCII or binary communication mode

Data Compression Algorithm

Scan data is compressed in binary mode to decrease communication time. In the compression scheme, each pixel value is compared to the previous value. If the absolute value of the difference is less than 128, the next byte is a signed 8-bit integer indicating the difference from the previous value. If the difference is 128 or more, then 3 bytes are sent with the following meaning:

Byte 1: the value 0x80 as a flag that the next two bytes represent a pixel value rather than a difference.

Byte 2: the high-order bits for the 16-bit unsigned integer representing the value at this pixel.

Byte 3: the low-order bits for the 16-bit unsigned integer representing the value at this pixel.

After all scan bytes have been transferred, a checksum byte is sent.

DIY Spectrometer Conclusion

The DIY Spectrometer offers an easy an inexpensive means to experiment with spectroscopy. With a bit of work, it is possible to couple the spectrometer to a telescope and do astronomical spectroscopy. Considering the cost of using other spectrographs, one can hardly go wrong. It is possible to do excellent spectroscopy with no other equipment other than a computer. Just the cost of a single least expensive CCD camera is over twice the cost of the DIY Spectrometer. It is strongly suggested that if these spectrometers become available again on eBay, seriously consider getting one.

Chapter 5

ALPY 600 Mid-Resolution Spectroscopy

Introduction

The ALPY 600 is a clever device, but it can be confusing with all the parts and possible modes of operation. To be used effectively it is very important that the Modules be properly adjusted. The complete ALPY 600 spectrograph consists of three modules, The Basic, Guiding and Calibration Modules. The Guiding and Calibration Modules are optional. Because of the vast variety of possible configurations and use of the ALPY 600 only the Basic and Guiding Modules will be discussed in detail. When using the Guiding Module the Basic Module's slit plate is replaced with the 23 µm Reflective Slit from the Guiding Module package.

The Module

Basic Module

The Basic Module (~$760 US$) contains the grism (transmission grating and prism) and adjustable slit. The Basic Module can be used by itself or with one or two more ALPY modules to increase the functionality. A 1.25″ telescope coupler is provided for coupling just the Basic Module to a telescope. The Basic Module can also be used to determine the band pass characteristics of filters.

Guiding Module

A Guiding Module (~$850 US$) provides a means of getting the star on the slit and guiding for time exposures. A reflective 23 mm slit is included with the Guiding Module. This is used in place of the slit plate that comes with the Basic Module. When coupling the Guiding Module to a telescope a 2″ coupler is provided, as opposed to the Basic Module's 1.25″ coupler. Most 8 in. and larger SCT's have a 2″ back, but many come with just a 1.25″ visual back. There are several aftermarket 2″ telescope couplers available, however. OPT (http://www.optcorp.com) has several for different telescopes. They range in price from $40 to $130, but for the Celestron and Meade telescopes they are around $60. Orion Telescope also carries these 2″ telescope couplers.

Note: The connection for the guiding CCD camera uses a C- mount. Most CCD cameras use a T-thread so a C-Mount to T-thread adapter will need to be purchased. Be sure to order a female C-Mount to male T-thread adapter

Instead of the Guiding Module various DIY arrangements could be used, such as a star diagonal with hole in the mirror and a fiber optic interface. The star diagonal and fiber optic interface is similar to what was used for the DIY Spectrograph. A beam splitter or a flip mirror could also be used. Be aware while these may sound simple, the devil is in getting it right. Unless experimentation is something enjoyed, the Guiding Module works well and without any experimentation and frustration.

Calibration Module

A third module is a Calibration Module (~$870 US$). This Module is used to provide standard wavelengths for calibrating the spectrum's line profile. The Calibration Module contains an Ar/Ne/H lamp for wavelength calibration and a tungsten lamp for flats. Like the Guiding Module, this is rather expensive and there are a multitude of options that can be used for wavelength calibration that are considerably less expensive. When using the Calibration Module the 2″ telescope coupler is removed from the Guiding Module and used to couple the Calibration Module to the telescope. The Calibration Module is connected to the Guiding Module with four screws.

The three units total is about ~$2,500 US$. The prices will vary daily due to the French/US$ monetary exchange rates. These prices do not include the imaging and guiding CCD cameras. Various digital cameras can be used. The design allows starting less expensively with just the Basic Module and then adding the optional modules for a more sophisticated spectrograph system. The ALPY 600 modules can be purchased from Woodland Hills in California at http://telescopes.net/.

The ALPY 600 Modules each come with a .pdf User Manual. Refer to these manuals for more details on adjusting and using the modules. These manuals can be downloaded from the Shelyak web site. The following is a brief discussion of the modules.

Basic Module

Introduction

The Basic Module box has 11 items, two caps, a small Philips screwdriver, two hex wrenches, a CD ROM with documentation, a 1.25″ Telescope Coupler, an External Body that encloses the Core Element and Body, a Body that encloses the Core Element, (which contains the grism, lens, filter and slit plate), and a CCD Camera Adapter (Fig. 5.1).

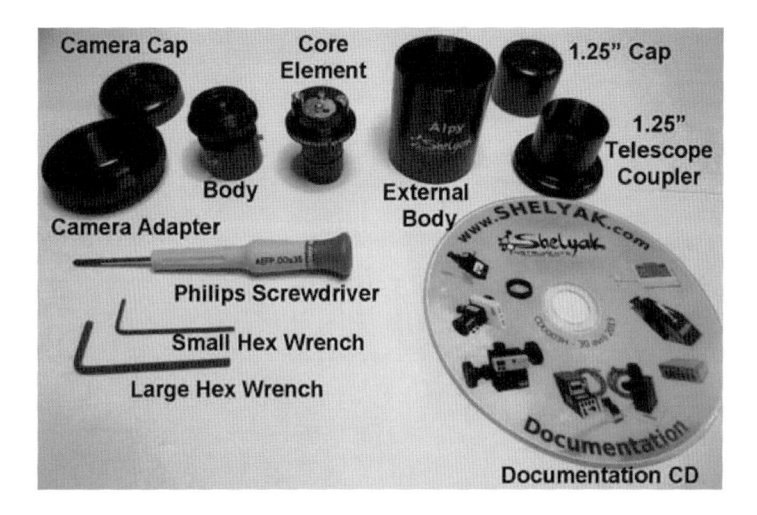

Fig. 5.1 In the basic module box

The Basic Module can be used with the Guiding Module or directly on a telescope using the 1.25″ Telescope Coupler. When using the Guiding Module, the 1.2″ Telescope Coupler is not used (Figs. 5.2 and 5.3).

Fig. 5.2 Basic module with CCD camera

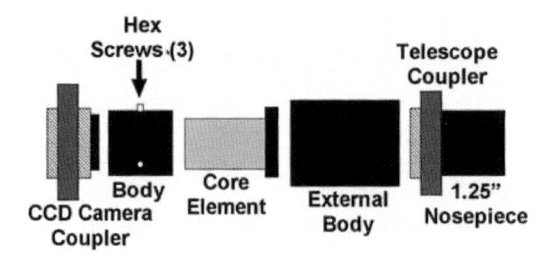

Fig. 5.3 Basic module elements

The Basic Module can be used as a slitless spectrograph or with several different slit configurations. The slit plate has multiple positions (25 μm hole; 25 μm, 50 μm, 100 μm or 300 μm slits; these are with a 5 μm tolerance) and with a clear position (3 mm hole) for a slitless mode. In slitless mode the ALPY 600 can be used similar to the Star Analyser, but at a higher resolution. The Basic Module can be used with a telephoto lens or telescope.

Basic Module

Note: When using the Guiding Module the slit plate is removed and a 23 μm reflective slit that comes with the Guiding Module is installed in its place (Fig. 5.4).

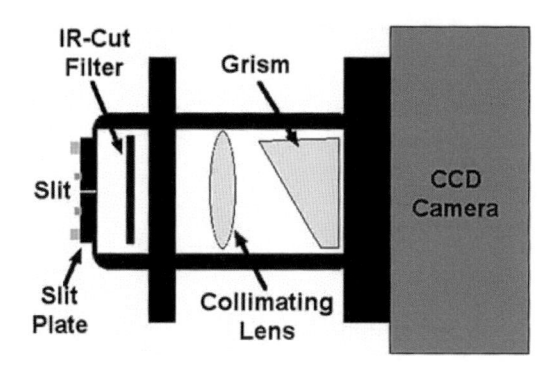

Fig. 5.4 Basic module core element X-ray

The ALPY 600 is designed so Basic Module can be used as a stand-alone Module. However, unless something like a fiber optic interface or the optional Guiding Module is used, it may prove very challenging. Shelyak suggests using a flip mirror or beam splitter between the telescope and Basic Module. Experience has shown unless one can see the star in the slit and track to keep the star in the slit, the whole operation becomes very frustrating. Using the Basic Module in slitless mode may work better however. One exercise that works well with just the Basic Module is bench-top experimentation.

Like the DIY spectrograph the Basic Module allows examination of the spectrum of various light sources, e.g., neon lamps, low pressure sodium lamps, CFL, laser pointers, etc., all without a telescope, and therefore no concern about finding a star, getting it into the slit and tracking it. In fact this is a suggested exercise to get familiar with the Basic Module and the spectrum processing software prior to taking it to the telescope. Everything learned from the bench-top experiments will make astronomical spectroscopy much easier.

ALPY Spectrum Imaging CCD Camera

Most any CCD camera can be used for the spectrum imaging, however a monochrome CCD camera is best. The most important factor is the back focus for the camera must be 19 mm (0.75″) or less. When using the Orion StarShoot G3 monochrome camera the back focus is 19 mm, but it was found that it was not possible to achieve best focus on the Basic Module if the locking ring is in place. CCD cameras with a shorter back focus should not have this problem. This is because the focus cannot get close enough to the camera. Removing the locking ring allows the focus to be achieved, but then no means of locking down the focus. There is about 0.040″ lack of travel with the focusing ring installed. To solve this there is a small lip on the camera coupler that is 0.165″ high. By milling this to a height of 0.100″ the problem was solved. After discussions with Shelyak it was determined that the next Basic Modules will have this problem corrected and no machining should be needed (Fig. 5.5).

Fig. 5.5 CCD camera coupler before and after

The size of the pixels of a CCD camera will determine the width of the spectrum window. The ALPY 600 tested for the spectrum window used a continuous spectrum and hydrogen gas discharge tube wavelength calibration. The results showed a spectrum window of 3700–7200 Å. The CCD camera used was an Orion StarShoot G3 monochrome camera. This has about an 8 μm pixel.

The following table shows the relationship CCD pixel size and spectrum window limits (Table 5.1).

Table 5.1 Spectrum window versus CCD pixel size

Pixel size (μm)	λ minimum	λ maximum
4	4560 Å	6055 Å
5	4293 Å	6322 Å
6	4026 Å	6589 Å
7	3759 Å	6855 Å
8	3650 Å	7122 Å
9	3650 Å	7389 Å
10	3650 Å	7656 Å
11	3650 Å	7923 Å
12	3650 Å	8190 Å
13	3650 Å	8457 Å
14	3650 Å	8500 Å

Basic Module 131

Adjusting the Basic Module

It is assumed for the following information that just the Basic Module is used (Fig. 5.6).

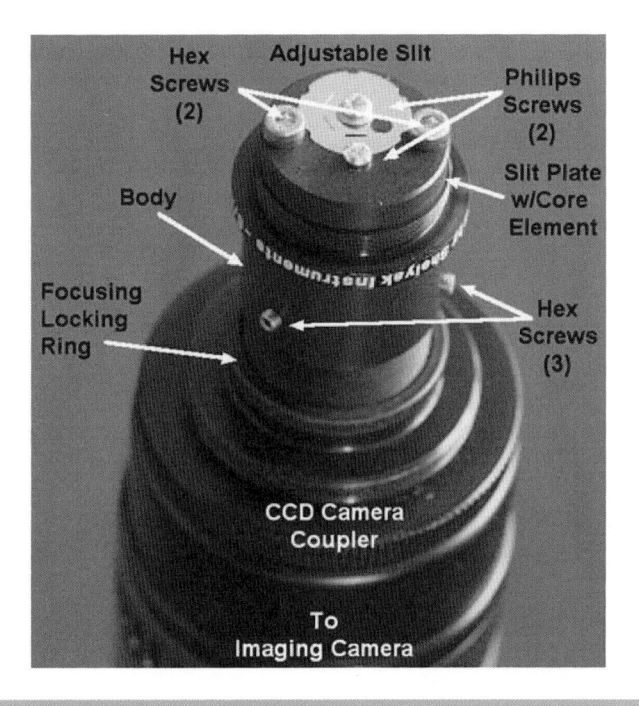

Fig. 5.6 Basic module adjustments

Note: The Basic Unit Core Element has an infrared cut filter that cuts off photons above 7200 Å. The purpose is to limit the spectrum image to just the first order spectrum. Do not remove this filter unless work in the region above 7200 Å is desired.

Slit Adjustment

There are several adjustments that need to be done to the Basic Module. The first is to select the slit to be used. Using a Philips screwdriver loosen the two screws holding the slit. Do not loosen the two hex screws. The hex screws hold the slit plate assembly. Rotate the slit plate until the desired slit, to start select the 25 μm slit, is at the center and then tighten the two Philips screws.

Core Element Focusing

Next install the CCD Camera to the CCD Camera Coupler/Body. The focusing can be accomplished by exposing the Basic Module slit end to light so that a spectrum is formed. It is suggested using a light source that produces some spectral lines rather than a continuous spectrum. Focusing is achieved by rotating the Body with the Core Element inside. Once a good focus is achieved the locking ring can lock the position. While rotating the Body and Core Element, the spectrum orientation will be rotated too. Once the focus is done using the small hex wrench loosen the three hex screws on the body. This will allow the Core Element to be rotated to provide the proper spectrum orientation without changing the focus.

Image Orientation

It is very important to have the spectrum image in the CCD camera horizontal and oriented such that the shorter wavelengths (blue) are to the left and the longer wavelengths (red) are to the right. If a color camera is used this is obvious and the CCD camera can be rotated to the correct position. The big problem is most likely a monochrome camera will be used and therefore one cannot see the color. Some ways to check the orientation can be done using laser pointers. Shine the laser pointer on a piece of white paper in front of the slit. A red laser will produce a spectrum line to the right. A blue/violet laser with produce a spectrum to the left. A neon spectrum will be concentrated at the right (Fig. 5.7).

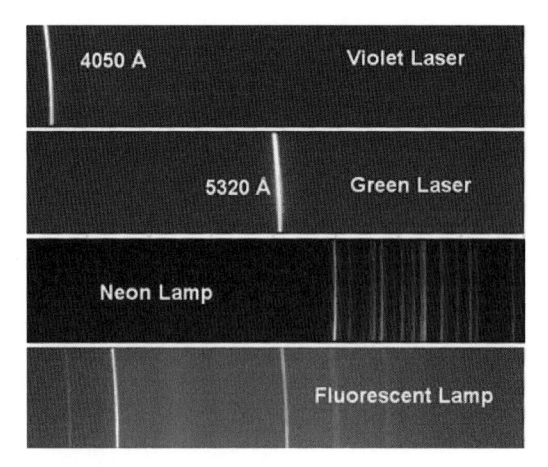

Fig. 5.7 Spectrum orientation

Guiding Module

Introduction

For astronomical spectroscope it is essential to have a means to guide the system to keep the star in the slit. The Guiding Module is an excellent means to accomplish that (Fig. 5.8).

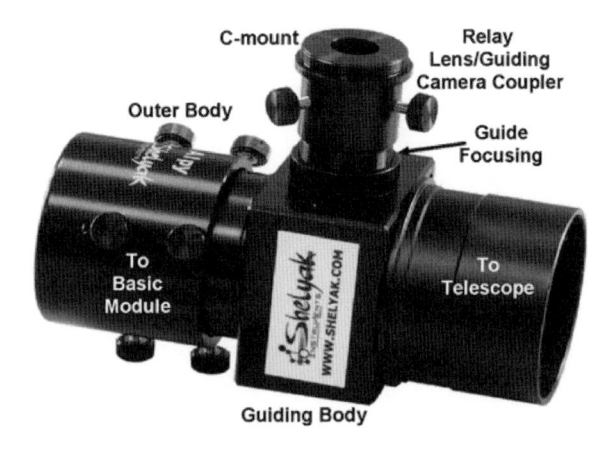

Fig. 5.8 Guiding module

Trying to use the ALPY 600 without the Guiding Module can be a big challenge. There is no easy way to find and get the star into the slit or to help keep it there without the Guiding Module. While rather expensive the Guiding Module can be added to allow easier guiding and centering of the star image on the slit.

The Guiding Module has six major parts, the Guiding Body with the Guiding Mirror, the Telescope Coupler, the Inner Body (where the Basic Module goes), the Outer Body (which encloses the Inner Body), the Relay Lens Holder (with focusing threads and locking ring) and the Relay Lens/Guiding Camera Coupler. The 23 μm reflective slit comes with the Guiding Module, but is not used in it. The Telescope Coupler from the Basic Unit is used in conjunction with the Guiding Module's Telescope Coupler to couple to a telescope.

The Guiding module contains a guiding mirror and relay lens. The relay lens allows the Reflective Slit to be seen, via the mirror, in the guiding camera. A fine focus is provided on the guiding port.

The Guiding Module includes a separate 23 μm Reflective Slit that replaces the Basic Module's slit plate. The Telescope Coupler from the Basic Module is used to connect the spectrograph to a telescope. The Basic Module's External Body is not used when using the Guiding Module (Fig. 5.9).

134   5 ALPY 600 Mid-Resolution Spectroscopy

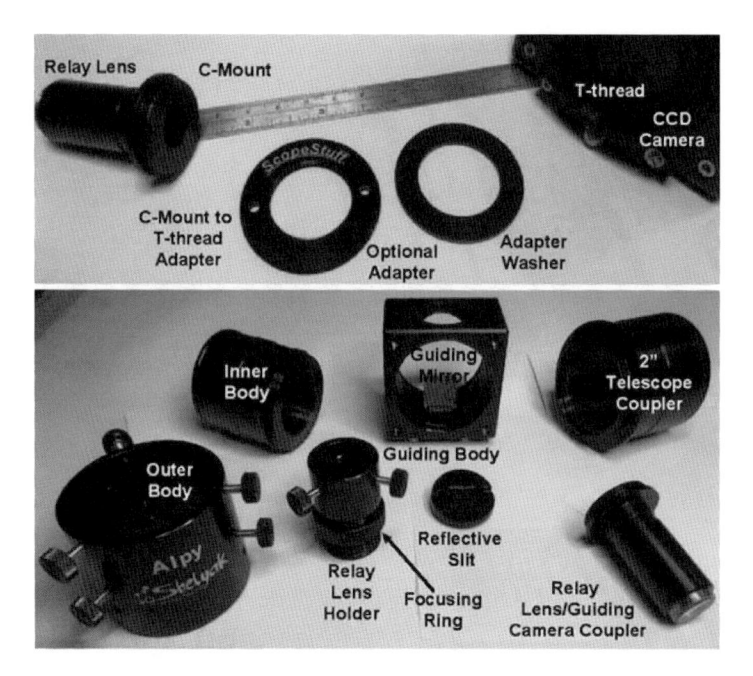

Fig. 5.9 Guiding module elements

For guiding a CCD camera with a computer must be used. The CCD guiding camera most likely with have a T-thread type port. The guiding port on the Guiding Module has a C-Mount type thread so a C-Mount to T-thread adapter must be purchased and used. ScopeStuff has one for $24 (#T2CA) (Fig. 5.10). http://www.scopestuff.com.

Fig. 5.10 Coupler with extra C-mount to T-thread adapter

To couple the Guiding Module to a telescope a 2″ telescope coupler or other means to accept a 2″ device must be on the telescope. For the tests of the ALPY 600 a Crayford focuser with a 2″ coupler was used on an 8″ LX90 telescope.

The Guiding Module provides an excellent way to guide for astronomical spectroscopy, but because the Guiding Module is rather expensive, more than the Basic Module, it may be wise to investigate other means of guiding (Fig. 5.11).

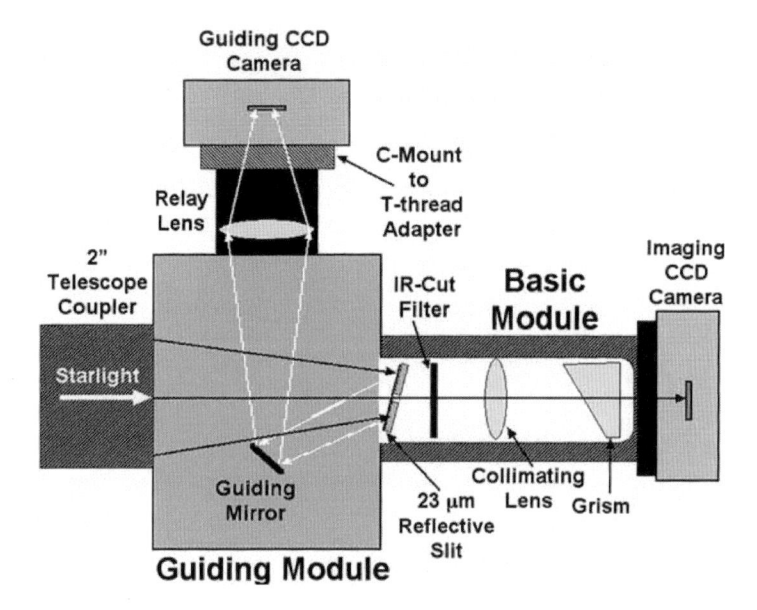

Fig. 5.11 Guiding module and core element X-ray

Adjusting the Guiding Module

When using the Guiding Module there is a 23 μm Reflective Slit that replaces the slit plate on the Basic Module. The Reflective Slit does not go in the Guiding Module. The slit plate on the Basic Module is removed by removing the two hex screws on it. The Reflective Slit that comes with the Guiding Module is then put in place of the slit plate and secured with the two hex screws. After the Reflective Slit has been installed the focusing and spectrum image orientation should be checked and adjusted as noted in the above section on the Basic Module adjustments.

The Outer Body in Fig. 5.9, with the six knobs, is where the Core Element goes after the slit plate on the Basic Module has been replaced with the Reflective Slit. Once the Basic Module has been coupled to the Guiding Module, the Reflective Slit must be adjusted for proper orientation. It will be noted that unlike the slit plate, the Reflective Slit is tilted. This is to reflect the star image to the mirror in the Guiding Module.

To adjust the orientation of the Reflective Slit, remove the Relay Lens assembly and look into the port, see below. Then with the six knobs loosened on the Outer Body, rotate the complete Basic Module until the Reflective Slit is reflecting light down. This is with the closest part of the slit at the top. Once in the proper position the six knobs should be tightened to prevent further rotation of the Basic Module (Fig. 5.12).

Fig. 5.12 Guiding port and reflective slit orientation

The guiding camera is focused on the Reflective Slit by loosening the locking ring and two knobs on the Relay Lens/Guiding Camera Coupler. The Relay Lens/Guiding Camera Coupler is then rotated screwing it in or out to achieve focus. Once focus on the slit has been achieved the locking ring should be tightened to prevent further rotation. The slit image should be horizontal. The image can be made horizontal by loosening the two knobs on the Relay Lens/Guiding Camera Coupler and rotating just the guiding camera. Once in the proper orientation the two knobs should be tightened to prevent further rotation (Fig. 5.13).

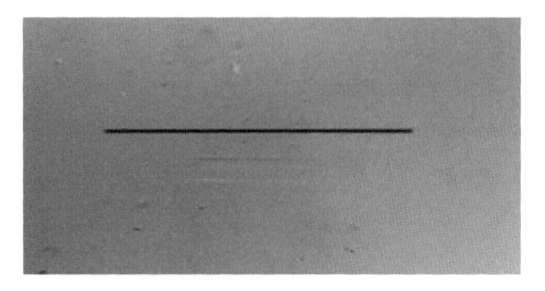

Fig. 5.13 Guiding port slit focus and orientation

Telescope Couplers

The Basic Module uses a 1.25″ Telescope Coupler. Because of the added weight and leverage of the Guiding Module, the 1.25″ coupling to the telescope is not sufficient. This is why the Guiding Module uses a 2″ coupler to the telescope (Fig. 5.14).

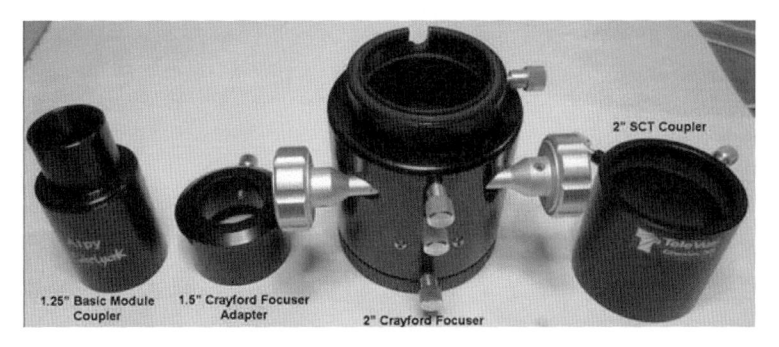

Fig. 5.14 Basic module 1.25″, Crayford focuser and 2″ SCT couplers

But wait, because the telescope port of the Guiding Module uses T-thread, the 2″ Telescope Couple can be removed and the 1.25″ Telescope Coupler from the Basic Module screwed into the Guiding Module. While this is NOT recommended, it will work. For best stability use the 2″ Telescope Coupler (Fig. 5.15).

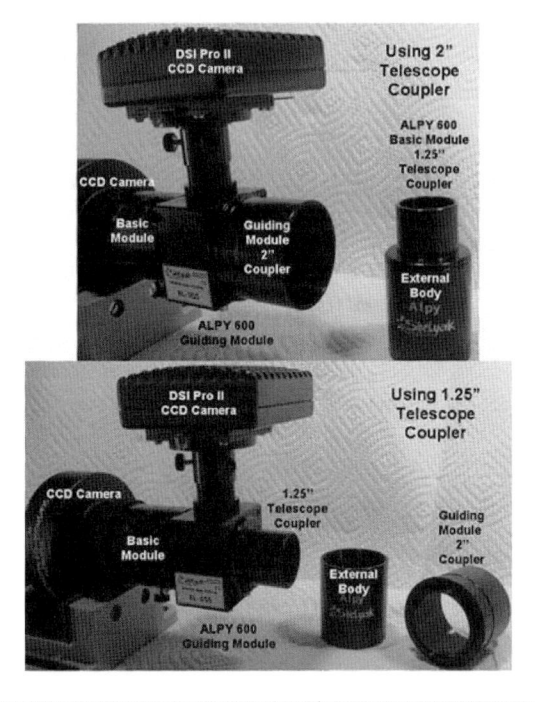

Fig. 5.15 Using the basic module 1.25″ telescope coupler

Calibration Module

Introduction

Because the Calibration Module was not tested, it will only be discussed briefly. The optional Calibration Module provides an easy was to get calibration spectra. The downside is that the Calibration Module costs $110 more than the Basic Module. The Calibration Module has neon/argon and tungsten lamps to provide known spectral lines for wavelength calibration and flat fields. There are multiple other means available to provide calibration for the ALPY that are considerably cheaper, but not as elegant (Figs. 5.16 and 5.17).

Fig. 5.16 Calibration module, credit Shelyak instruments

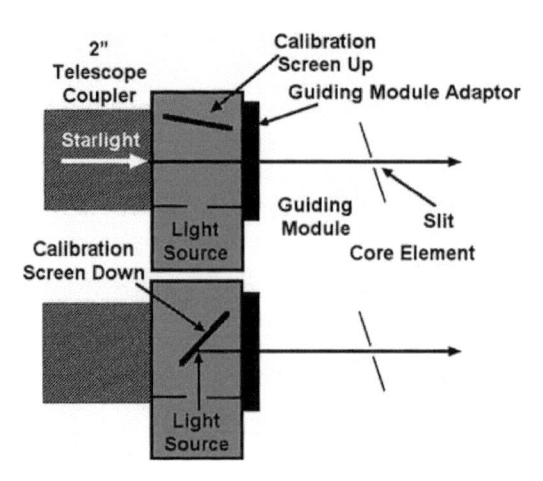

Fig. 5.17 Calibration module X-ray

Complete Assembly

The complete assembly consists of the Basic ALPY 600, Guiding and Calibration Modules (Figs. 5.18 and 5.19).

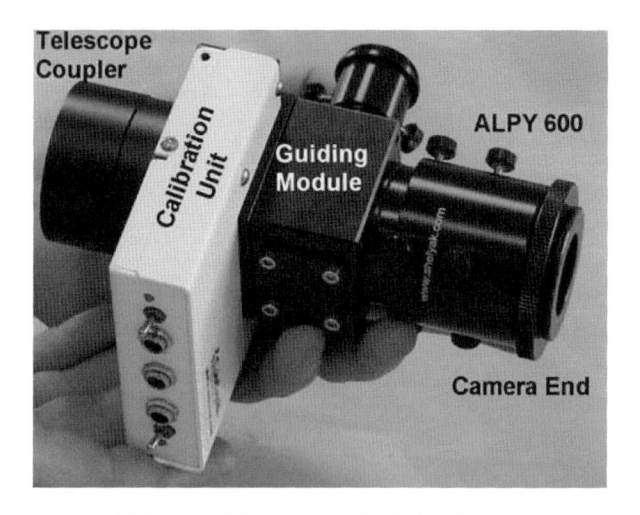

Fig. 5.18 ALPY 600 spectrograph assembly, credit Shelyak instruments

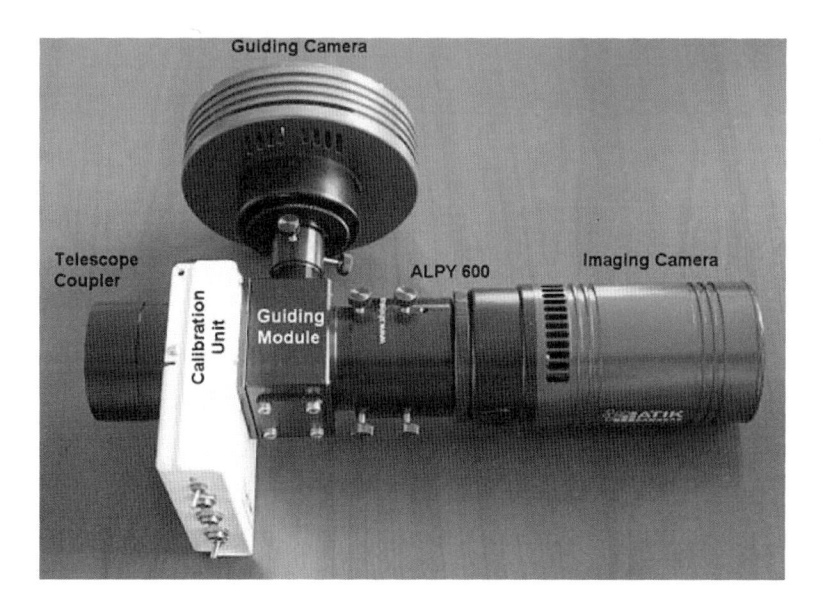

Fig. 5.19 ALPY 600 spectrograph with cameras, credit Shelyak instruments

Taking the Spectra with an ALPY 600

It is suggested that experimentation begin indoors on the bench before connecting the ALPY 600 to a telescope. This applies to use of the Basic Modules by itself and in combination with the Guiding Module. The Basic Module will need some adjustments before use and when the Guiding Module is added, more adjusts must be made. These adjustments are discussed above. Most of these are best done indoors in the light and on the bench. This will also help getting familiar with the various parts of the spectrograph. Making adjustments in the dark at the telescope is not recommended and can be very challenging and frustrating.

There are several modes of operation possible with the ALPY 600 spectrograph.

Basic Module on the Bench

Color spectra were obtained by exposing the Core Element to fluorescent and neon lamps and then taking images with a digital camera held to the end of the Core Element (Fig. 5.20).

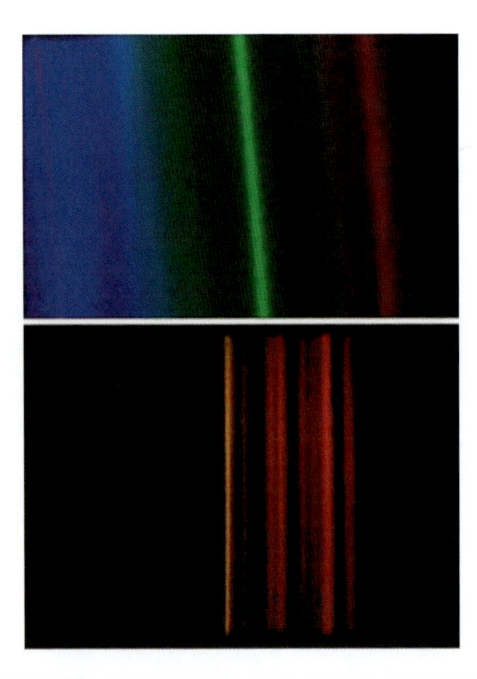

Fig. 5.20 Basic module fluorescent and neon lamp spectra

Taking the Spectra with an ALPY 600   141

The Basic Module was next connected to a CCD camera and fixed in a vise. A vise or other means of securing the assembly is recommended. A neon lamp was then used to produce a neon spectrum. Ambient light from a fluorescent light was also used to produce a spectrum of mercury. A violet laser was pointed at a white paper sheet in front of the slit plate and a spectrum produced. Similarly a green laser was used. (See Figs. 5.7 and 5.21).

Fig. 5.21 Basic module with neon lamp

Filter Band Pass Determination

The ALPY 600 Basic Module can be used to easily determine the wavelength characteristics of filters. The Basic Module was configured as if it was to be used with a telescope, but instead is setup on the bench. The filter to be tested is screwed into the 1.25″ nosepiece. The nosepiece has threads that accept standard filter cells. A diffused continuous spectrum light source, in this case an incandescent flashlight, was used to provide the light. A hydrogen alpha band pass filter was tested (Fig. 5.22).

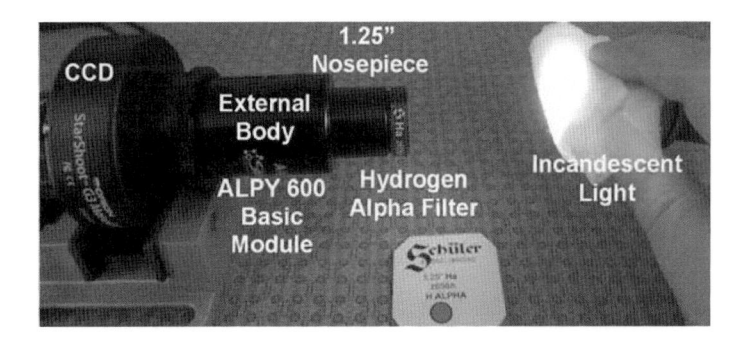

Fig. 5.22 Filter band pass determination setup

A 0.05 second exposure was used to produce a spectrum. The spectrum was then processed in RSpec to produce a line profile of the spectrum. A hydrogen gas discharge tube was used to calibrate the spectrum (Fig. 5.23).

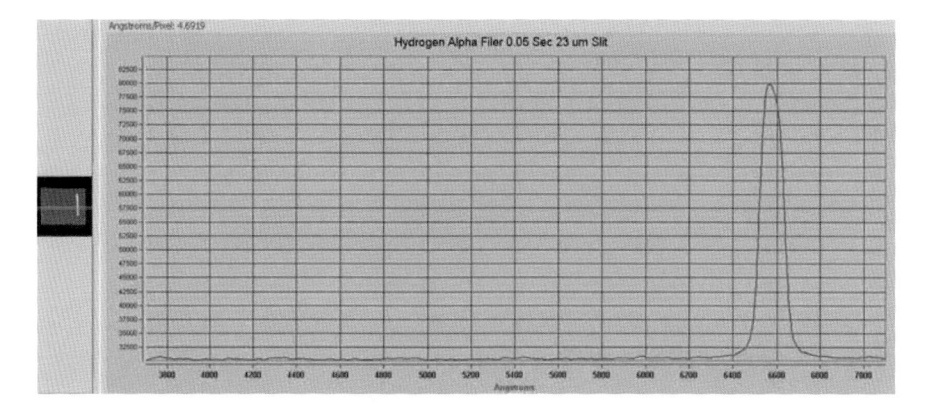

Fig. 5.23 Hydrogen alpha filter line profile

Basic Module on Telescope

Using the Basic Module by itself on a telescope is not recommended as it will be nearly impossible to get a star's light into the slit and keep it there. The following was done with a telescope on the bench to check out a means of using a neon ring and low-pressure sodium light in front of the telescope for wavelength calibration (Fig. 5.24).

Fig. 5.24 Basic module on telescope

Basic Visual Mode

There are two ways to use the Visual Mode. First is to use the core element as a hand held spectroscope. The second way is as an eyepiece with a telescope. Either of these methods can be in a slitless mode, similar to the Star Analyser, or with one of the available slits on the slit plate. The core element with the slit and grim assemble is what produces the spectrum. Everything else is an added feature. To use it visually without a telescope the slit end is held up to a light with the slit vertical and the spectrum viewed visually at the imaging end. For use visually with a telescope the core element is inserted into the telescope coupler and 1.25" nosepiece. The nosepiece can then be inserted into a telescope or star diagonal.

Basic CCD Camera Mode

Like the Basic Visual Mode this mode can be either in a slitless mode or with one of the available slits on the slit plate. While it can be used without a telescope for bench experiments using a CCD camera, for astronomical spectroscopy a telescope or telephoto lens is needed. With this mode the Basic Module with the core element inside is connected to a telescope and a CCD camera is coupled to the image port of the spectrograph.

Note: A means of finding the star, getting it in the slit and tracking the star must be made. As mentioned above there are several possible ways to do that, a star diagonal and fiber optic interface, beam splitter or flip mirror. Be aware, except for the slitless mode using any slit will be a challenge.

Guiding Mode on the Bench

For easy guiding the ALPY Guiding Module is used. The following Figure shows the Basic and Guiding Modules on the bench with a neon lamp for wavelength calibration (Fig. 5.25).

Fig. 5.25 Basic and guiding modules on bench

Basic and Guiding Modules on the Telescope

The following Figure shows the Basic and Guiding Modules on a telescope. An Orion StarShoot G3 monochrome CCD camera was used for spectrum imaging and a DSI Pro II monochrome CCD camera for guiding. The ALPY 600 was coupled to the telescope with a Crayford 2″ focuser (Fig. 5.26).

Fig. 5.26 Basic and guiding modules on telescope

Taking the Spectra with an ALPY 600

A preliminary observation was made of the bright star Vega using a 30-second exposure. An 8″ LX-90 F/10 telescope was used with the Basic and Guiding Modules with the 23 μm slit. The night was warm ~90° F, and it was found the spectrum was tilted slightly. The tilt was corrected with software. A non-linear wavelength-calibration was done.

A Dell Latitude computer with Windows XP Professional was used. Two imaging programs were used, Envisage for the DSI Pro II CCD camera (for guiding) and Orion Camera Studio for the Orion StarShoot G3 CCD camera (for spectrum imaging). Both programs can run concurrently so only one computer is needed (Fig. 5.27).

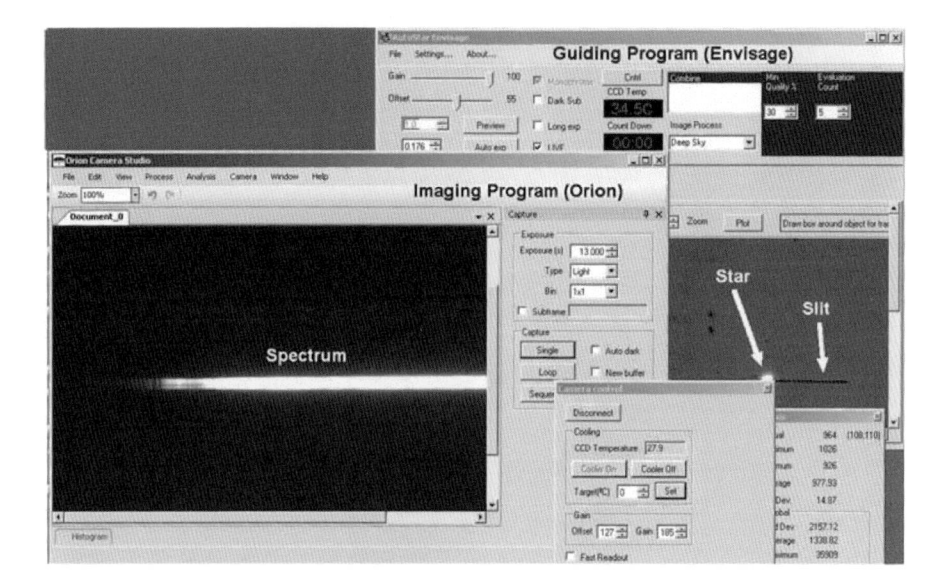

Fig. 5.27 Computer screen shot of observation

The basic procedure for processing the spectrum is:

1. Rotate the spectrum horizontally if necessary.
2. Subtract the background sky.
3. Wavelength-calibrate the line profile. A third order calculation was done using six points. The fit is very good as can be seen in Fig. 5.28.

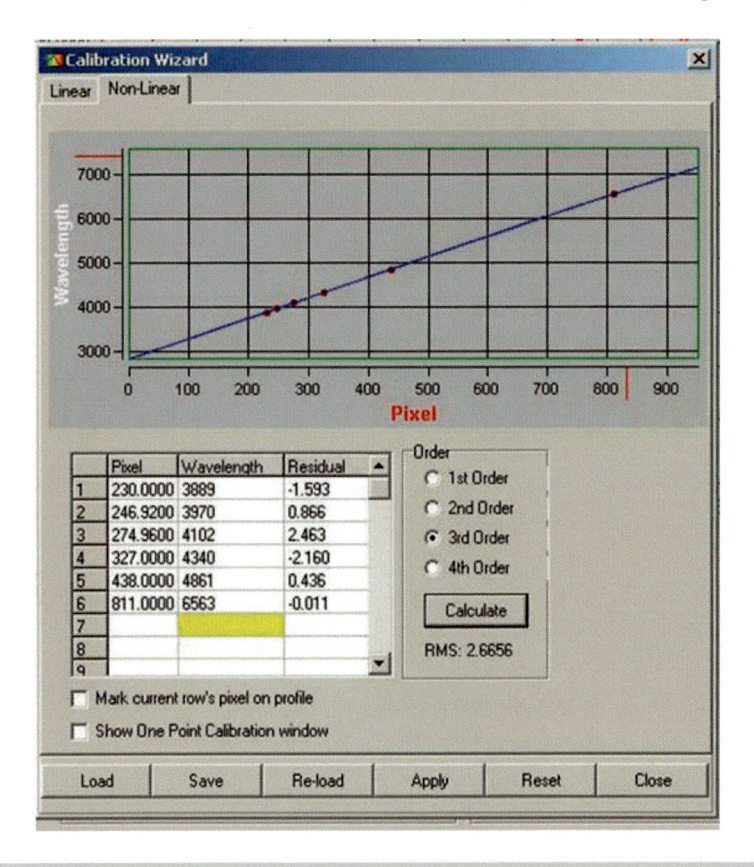

Fig. 5.28 Non-linear wavelength calibration (RSpec)

4. Finally the ends of the line profile were deleted to trim it up. It is important to use the same end points for trimming different line profiles and to trim the profile after the calibration. Otherwise previous calibrations may not work correctly.

The processing was done in the software program RSpec. The procedure is described in detail in the Software Chap. 7 on RSpec.

The following Figure shows the resulting calibrated line profile. The Hydrogen Balmer lines are very prominent (Fig. 5.29).

Taking the Spectra with an ALPY 600

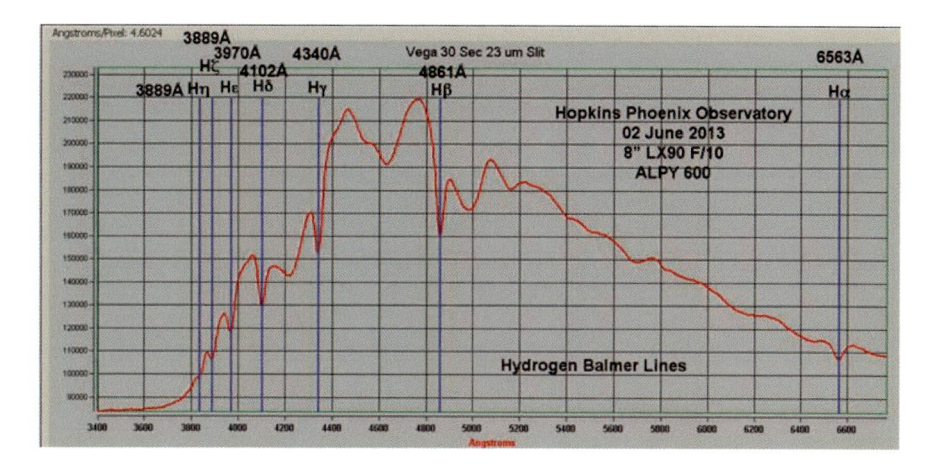

Fig. 5.29 Basic and guiding modules spectrum line profile

Many of the same issues connected with taking spectra with a Star Anaylser also apply to the ALPY 600. This is particularly true when using the ALPY 600 in a slitless mode. When using a slit the game changes some. In the slit mode the slit is fixed. What that means is once the unit it calibrated and not changed, the calibration should be close for all future spectra. Slight calibration adjustments may be needed, but the calibration should be close enough to easily identify some lines to refine the calibration. This is not true when in the slitless mode, however. Because the star position will determine where the wavelengths fall, it will not be possible to get the star in the exact same position each time, thus the calibrations must be redone for each image.

Calibration Images

As mentioned earlier the Calibration Module provides an easy way to take calibration images. The downside of using it is it is more expensive than the Basic ALPY module. There are easier and less expensive means to wavelength calibrate the ALPY. On the bench, a neon lamp calibrator can be constructed for just a couple of dollars. A low-pressure sodium lamp can also be used for calibration of the area around the sodium D lines. Once calibrated, the calibration should be good for future spectra unless the Basic ALPY module is disturbed. For use on a telescope an inexpensive neon ring can be constructed and mounted on the front of the telescope.

Neon Ring

When mounted on the telescope the ALPY can be calibrated with an inexpensive and easily constructed neon ring mounted on the front of the telescope. This provides an excellent means to check the wavelength calibration. The one shown in

Fig. 5.30 on an 8″ telescope is similar to the one described in detail in Chap. 6. A 1.0 second exposure provides a good neon spectrum. This can be turned on (for a calibration spectrum) and off (for a stellar spectrum) remotely.

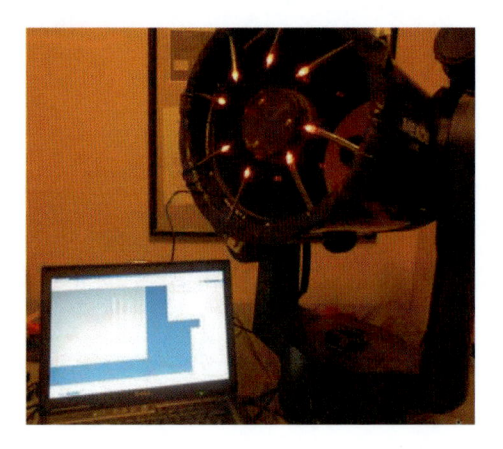

Fig. 5.30 Neon ring calibration

Low-Pressure Sodium Light

To provide a calibration spectrum for the sodium D line region a low-pressure sodium lamp can be held in front of the telescope. Details for constructing one of these lamps are given in Chap. 6 (see Figs. 6.16, 6.17, 6.18 and 6.19).

The following is a spectrum of sodium using the low-pressure sodium lamp held in front of the telescope. The spectra were produced with the neon ring and low-pressure sodium lamp both with a 1.0 second exposure (Fig. 5.31).

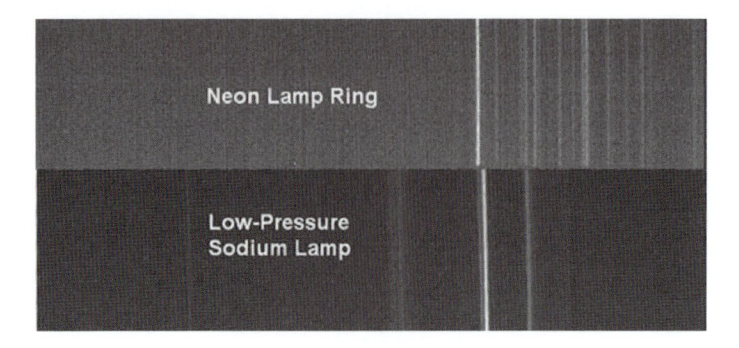

Fig. 5.31 Neon ring and low-pressure sodium spectrum

The following Figure is a neon calibration using a neon lamp and the 23 μm Reflective Slit. The blue lines are the standard lines for neon. The red profile is the line profile of the neon lamp spectrum. Because of the low-resolution of the ALPY 600 some of the actual neon lines are merged. Such is the case for the N1 (5852.8 Å) and N2 (5881.9 Å) standard lines. They merge in the line profile into a wider line center at about 5868 Å. Because there are some small amounts of other gases in the neon lamp some of the lines are not neon and some are a combination of the neon and other gases (Fig. 5.32).

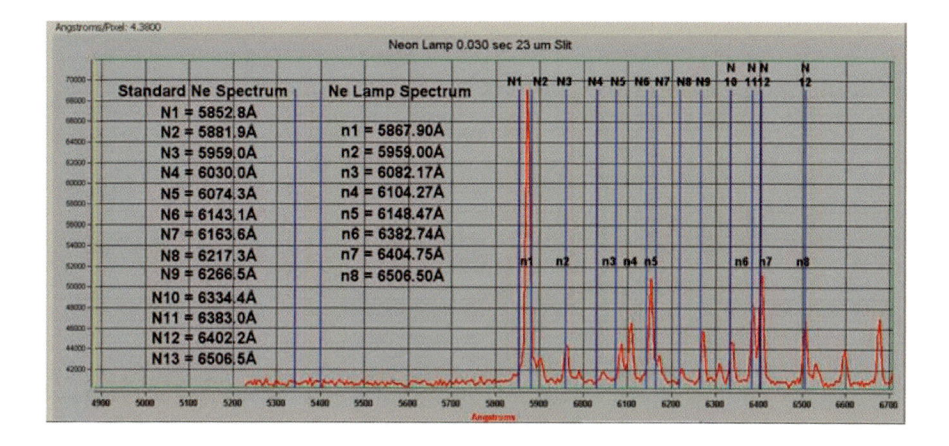

Fig. 5.32 Neon calibration

ALPY 600 Fiber Optic Interface

Using the Basic Module by itself on a telescope will not work well. The reason is the star or object of interest cannot be easily seen and put on the slit. Using a flip mirror will work, but is very challenging. Once the mirror is flipped up there is no indication if the star remains on the slit. Unless the telescope is near perfectly aligned the star will drift. If a beam splitter is used one can still not see the slit. Beam splitters have been used successfully, however. Using a star diagonal with a fiber optic interface is an inexpensive way to be able to acquire and track a star. The alternative is the use the ALPY Guiding Module. This is an excellent way to guide, but expensive ($850).

Shelyak offers a fiber optic coupler (~$120 US$, SE0131) for the Basic Module that replaces the slit plate. If this is used a small fiber is required as that will determine the resolution. The plastic fiber mentioned earlier is 1,000 μm and would not work well with this arrangement. Shelyak offers a high quality fiber optic cable (20 m and 200 μm) for $810 and for a telescope couple/guider for ~$3,800 US$. This route is rather expensive. For serious work it is suggested either some kind of fiber optic interface be used or adding the Guiding Module.

DIY ALPY Fiber Optic Interface

A fiber optic interface for the ALPY Basic Module can be constructed for under $50. Most of this is described in Chap. 4. The difference is the ALPY Basic Module's Core Element is used for the spectrograph rather than the DIY spectrograph. To couple the fiber optic cable to the Core Element a 1.5″ diameter 1.0″ thick disk was used. This can be made of wood, plastic or aluminum. For this experiment two 0.5″ thick aluminum plate disks were used. A 12' TOSLINK 1.0 mm fiber optic cable connects the star diagonal to the Core Element fiber optic coupler. A 0.275″ (lettered drill J) diameter hole was drilled in the coupler for the TOSLINK connector. The connector is plastic and has some small keys on it. By using a sharp knife the small keys on the side can be trimmed so the connector fits tightly into the hole (Figs. 5.33 and 5.34).

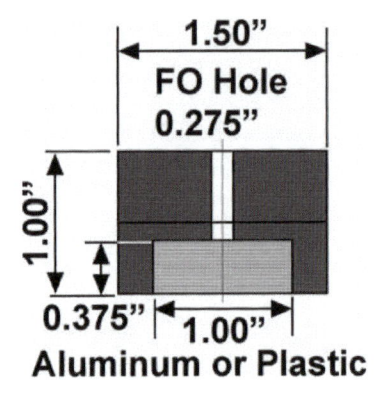

Fig. 5.33 Basic module fiber optic coupler

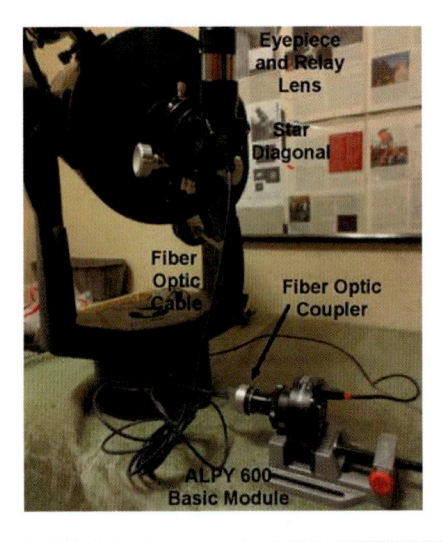

Fig. 5.34 Basic module fiber optic telescope interface

Chapter 6

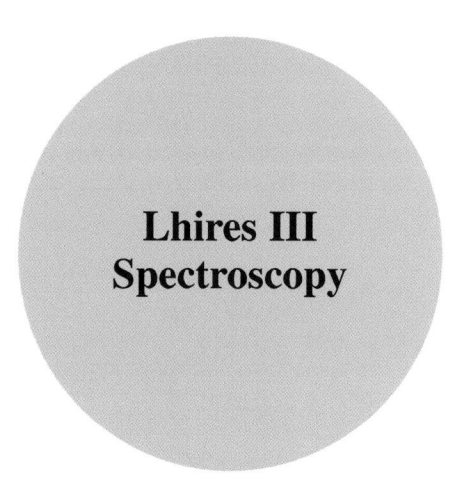

Lhires III Spectroscopy

Introduction

While there are several commercial medium and high-resolution spectrographs on the market, the most popular unit is the Lhires III by Shelyak. Shelyak also markets the ALPY 600 (medium-resolution), LISA (medium-resolution) and the high-end eShel spectrographs, but only the Lhires III will be discussed in detail in this Chapter. There are several different gratings available for the Lhires III. The Lhires III with the stock 2,400 l/mm and optional 600 l/mm gratings will be discussed.

Digital Cameras

While any digital camera can be used with the Lhires III to acquire a spectrum image, even DSLR and other digital cameras, the preferred camera is a monochrome CCD camera with a 16-bit ADC. Bigger chips will show more of the spectrum, but come at a financial and physical space cost. In addition, the bigger the chip, the longer the download time. Normally the download time is not much of a consideration, but when setting up and experimenting, it can be maddening.

The Lhires III Spectrograph

The Lhires III is a Littrow type spectrograph with a folder light path. This makes the unit compact yet still high-resolution. This design is ideal for use with most amateur telescopes. It is optimized for a 12″ F/10 telescope, but telescopes from 6″ to 16″ and F/5 to F/15 will work fine. The most important consideration is that the telescope and mount be sturdy enough to hold the spectrograph without flexing (Figs. 6.1 and 6.2).

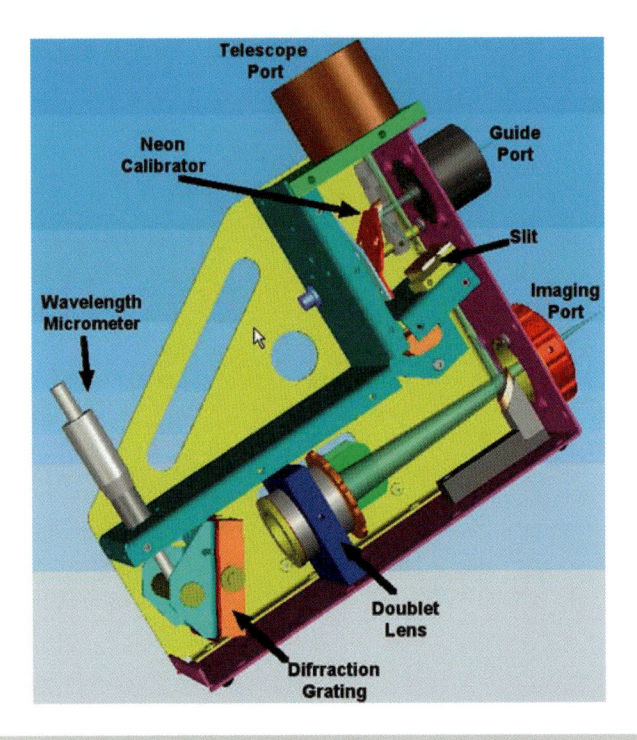

Fig. 6.1 Lhires III spectrograph X-ray, credit: Shelyak instruments

The Lhires III Spectrograph

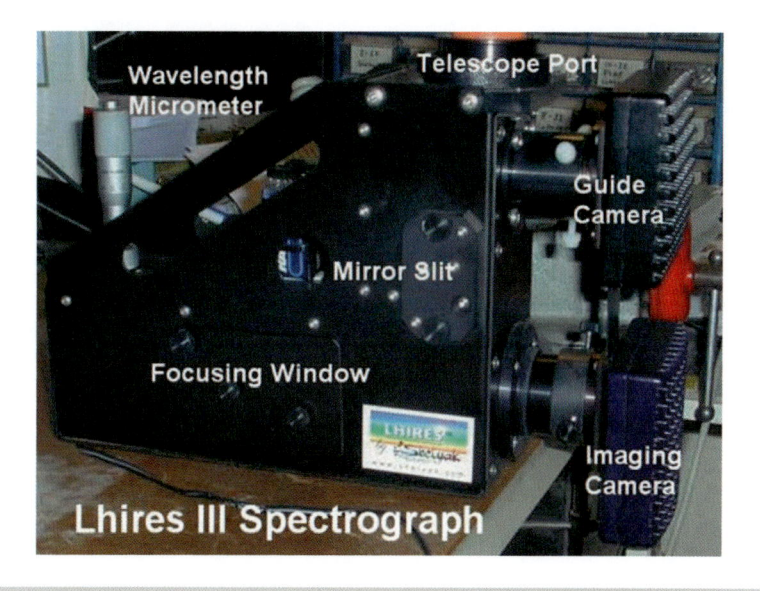

Fig. 6.2 Lhires III spectrograph on the bench

The Lhires III should come pre-adjusted. If adjustments are needed, refer to the Lhires III User Manual. Typically the earlier Lhires III came with adjustable slits set at approximately 22 mm. The newer units have a fixed glass slit that has several selectable positions. When ordering a Lhires III, be sure to specify what CCD cameras you plan to use so the proper adapters can be supplied.

Gratings

The stock Lhires III comes with a 2,400 l/mm grating, but 1,200, 600, 300 and 150 l/mm are available. The spectrograph is factory adjusted for use with the 2,400 l/mm grating, but appears to work fine with the 600 l/mm grating without any readjustments. There are four screws on the grating end of the Lhires III. Removing these screws provides access to the grating for removal. All the gratings come mounted in a holder and the whole grating unit is removed. The micrometer should be screwed close to the top with a reading around 2,200. The reason for this is the micrometer lower shaft pushes on the spring-loaded roller to change the grating's position and thus wavelength window/wavelength coverage. Be careful not to move the micrometer up too far. The bottom of the micrometer has a small rubber cap that can be pushed off. When the grating is removed, check to see if it is still there. If it is missing, it will be inside the Lhires III housing. Find the cap and replace on the bottom of the micrometer before installing the grating unit (Fig. 6.3).

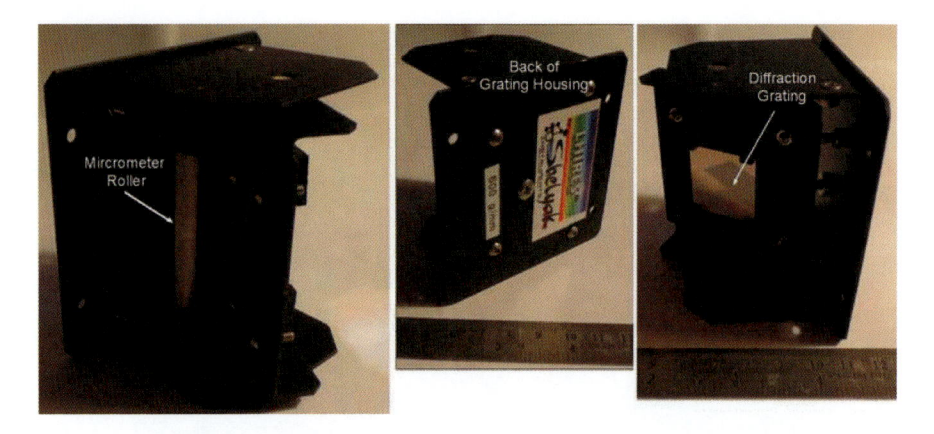

Fig. 6.3 Optional 600 l/mm grating unit

Lhires III Modifications and Adjustments

The Lhires III should be usable as it comes from the factory, however there are some things that can enhance it. First, the neon calibrator is designed to work with 12 VDC. The active device inside can work with voltages from 10 to 18 VDC. This means two 9-V batteries could be used. The down side of using batteries is replacing or recharging the batteries.

The second thing is some of the early Lhires III bonded the gratings to the holder. These caused the Lhires III to have a bad temperature response. The result is even a small temperature change can cause the spectrum to change focus. If you have a bonded grating rather than one held by clamps, contact Shelyak and see if they will replace it for you. The gratings with the mechanical clamps are less temperature sensitive.

The Lhires III leaks light. In a totally dark environment this is no problem, but if any light is showing, even a red light, the spectrum can be compromised. The grating end seems to be the worst leak. The solution is to duct tape all the seams.

Perhaps the most frustrating and poorest design of the Lhires III is the focusing access plates on each side. These plates are a pain to remove and replace in the light and a nightmare in the dark. A better solution is to use duct tape. Details are explained later in the focusing section.

Micrometer Calibration

The micrometer used to adjust the wavelength window of the grating is like most micrometers. The major marks go from 0 to 25. The minor (rotating) marks go from

The Lhires III Spectrograph

01 to 50. Two turns of the micrometer will move one major mark. Rotating the micrometer CCW part of one turn past the major mark at 17 and stopping on 45 will produce a setting of 1,745. If it were part of the second turn the setting would be 1,795. As will be discussed later it is important that turning the micrometer CCW (up) take the grating to longer (red direction) wavelengths.

It is suggested that a calibration of micrometer setting vs. wavelength be made for each grating to be used. This is only an approximate calibration, but should allow the micrometer to be set so the center of the image is close to the corresponding wavelength. It also allows the user to become more familiar with the Lhires III.

Use the neon calibrator to help with the calibration. Identifying the lines and their wavelengths is no easy task, but will be an excellent bench exercise. It is a bit like solving a mystery and as you solve more the more confident you become. Part of the problem is the neon lamps have some other gases in them that produce lines that are not neon lines. Still the major neon lines can be identified. This experience can be used later when identifying lines of a stellar spectrum line profile.

Other sources can be used to help, e.g., Blue/Violet, Green and Red laser pointers. These can provide single line references. Be aware, these lasers do not all have the same wavelengths, i.e., depending on which red laser is used, the red laser pointer could have one of several wavelengths. Make sure you know which you are using. The green and blue seem pretty standard, however (Figs. 6.4, 6.5 and 6.6), (Table 6.1).

A low-pressure sodium lamp is excellent for identifying the sodium D lines at 5889.950 Å and 5895.924 Å

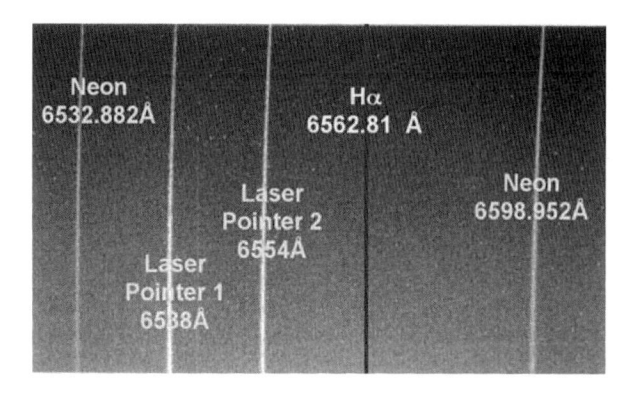

Fig. 6.4 Neon and laser pointer lines around hydrogen alpha

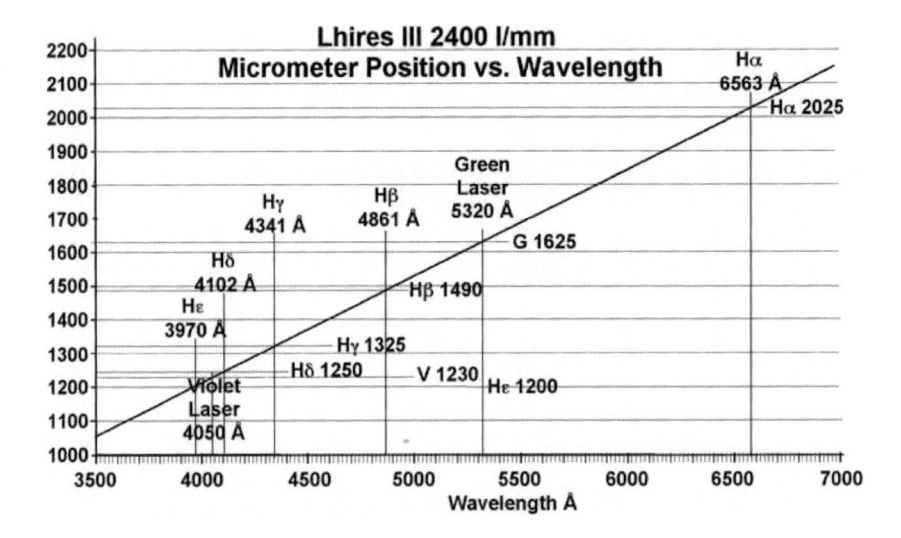

Fig. 6.5 Lhires III with 2,400 l/mm grating micrometer calibration

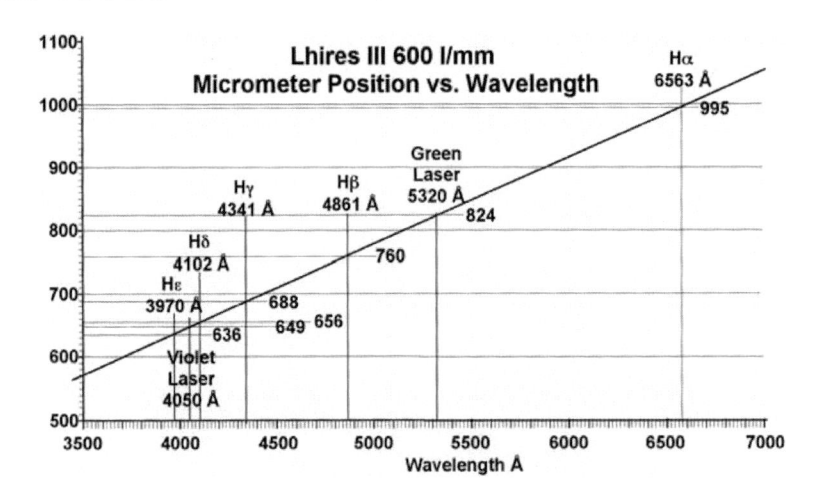

Fig. 6.6 Lhires III with 600 l/mm grating micrometer calibration

Table 6.1 Laser pointer wavelengths

Laser pointer wavelengths	
Blue/Violet	4050 Å
Green	5320 Å
Red 1	6560 Å
Red 2	6571 Å
Red 3	6621 Å
He/Ne	6330 Å

Mounting the Lhires III on the Telescope

Because of the weight and leverage of the spectrograph and the desire for rigidity, the unit has 2″ (50 mm) SCT backend type fitting so it can be screwed directly to the back of the telescope. A 1.25″ adapter will not work properly. This is very important. Make sure the telescope is in balance with the spectrograph, cameras and cables attached. While the orientation is not critical and can be adjusted to produce a spectrum image of a select star with close companions, usually the preferred orientation is such that the star image will drift in RA along the slit opening. This way slight periodic RA error will just cause the spectrum image to stay in the slit and have more height, which is actually good (Fig. 6.7).

Fig. 6.7 Lhires III mounted on 12″ LX200 telescope

In the image above the Lhires III is connected to a Meade electrical focuser. It was found that the instrument was too heavy for the focuser and while the focuser could focus out, it would slip when trying to focus in. As a result the electronic focuser is no longer used at HPO.

Adjusting the Lhires III

The following adjustments are in addition to any factory adjustments and are adjustments needed for producing a spectrum image. Remember there are two different cameras used, a Guide camera and Imaging camera.

Guide Camera Orientation

While there is no critical orientation for the guide camera, usually it is oriented so the slit appears horizontal. First use an external light, such as daylight or a flashlight shone into the front of the telescope, to let the slit be seen. Rotate the camera until the slit is seen horizontally. This along with the focusing can easiest be done on the bench before mounting the spectrograph on the telescope. Once the spectrograph is mounted on the telescope then rotate the whole spectrograph so that RA drift is along the slit.

Image Camera Orientation

The orientation of the spectrum imaging camera is very important. There are two considerations. First the shorter (blue) wavelengths of the spectrum should be to the left and longer (red) wavelengths to the right of the image on the computer. Note that with high-resolution spectroscopy, the wavelength orientation is not obvious, even if a color camera is used. To test the wavelength orientation, use the neon calibrator to display some neon lines. Then turn the spectrograph's micrometer in, CW, toward lower numbers on the micrometer. This should move the shorter wavelength neon lines from left to right. If the lines move from right to left, then the camera should be rotated 180°. This orientation is the standard way to image a spectrum. Next, the spectrum should be as horizontal as possible. Rotate the camera until the above conditions are satisfied and lock it down. Note the neon and other calibrating lines in the Lhires III will appear slightly curved. This is normal. The important part is to get the star's spectrum's continuum horizontal. Most of the above can be done indoors on the bench.

Focusing

There are three different areas that need focusing, guide camera, spectrum, and telescope (star image).

Guiding Camera Focusing

It is important that the guiding camera be focused on the slit. This is accomplished by physically moving the guiding camera in and out of the guiding port. This is easiest done with the Lhires III off the telescope and on the bench. Pointing the telescope port toward a light or sky (not the Sun) should illuminate the slit so it can be focused on. Once focused it should not need refocusing unless the guide camera is removed and replaced (Fig. 6.8).

The Lhires III Spectrograph

Fig. 6.8 Guide camera slit image

Spectrum Focusing

The focus of the spectrum image in the imaging camera is affected by the position of the doublet lens, temperature and wavelength area. If the wavelength setting is changed significantly, the spectrum will need to be refocused. If the temperature changes significantly during the evening, the focus must be checked and adjusted if necessary. Note, whenever spectrum focusing is done, the wavelength positions will change slightly. Always take calibrating images after the focusing. This is very important. The calibration image can be taken before or after the star's spectrum image or even before and after and then averaged.

The image camera should be connected to the imaging port using the recommended spacers and couplers supplied with the spectrograph. No focusing adjustment should be done here, just lock down the adapter and camera. Refer to the Lhires III User Manual for more information.

Spectrum focusing is best accomplished using the internal neon calibrator and a 1.0 second exposure. The flip mirror lever/knob must be rotated to go from the star position to the calibrate position. The knob should be horizontal for the calibrate position. With 10–18 VDC applied, the Neon Calibrator power switch should be turned on and the green power indicator will glow (Fig. 6.9).

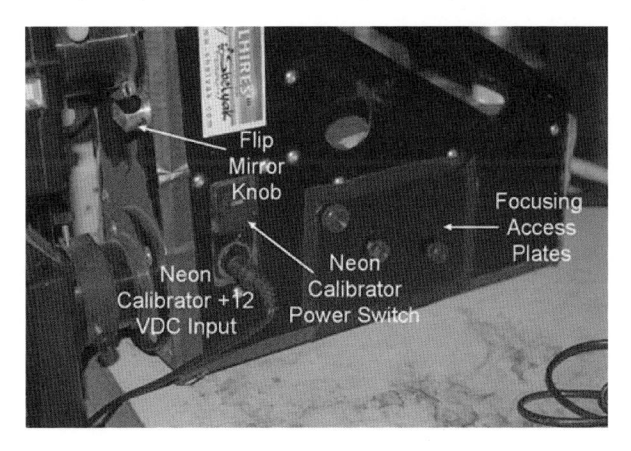

Fig. 6.9 Lhires III neon calibration and focusing side plates

Side Plates and Spectrum Focusing

The Lhires III has rectangular side plates on both sides of the unit that allow access to the doublet lens port. This is where the spectrum on the imaging camera is focused. These plates are cumbersome to remove and replace and not one of the better features of the Lhires III. While very low-tech, it has been found that rather than struggling with the side plates a simple solution is to remove them and use strips of duct tape to cover the focusing ports. These can be easily peeled partway back for access to the focusing ring and easily re-applied in the dark. The focusing ring can be stiff and may require fingers on each side to turn it. There are two nylon locking screws, one on each side of the focusing ring. These can be used to lock down the focusing ring once proper focus has been achieved. To adjust the focus the nylon screws must be loosened. Turning the focusing ring from just one side can at times be very difficult or impossible. That is why access to each side is provided. Both hands must be used to turn the ring (Fig. 6.10).

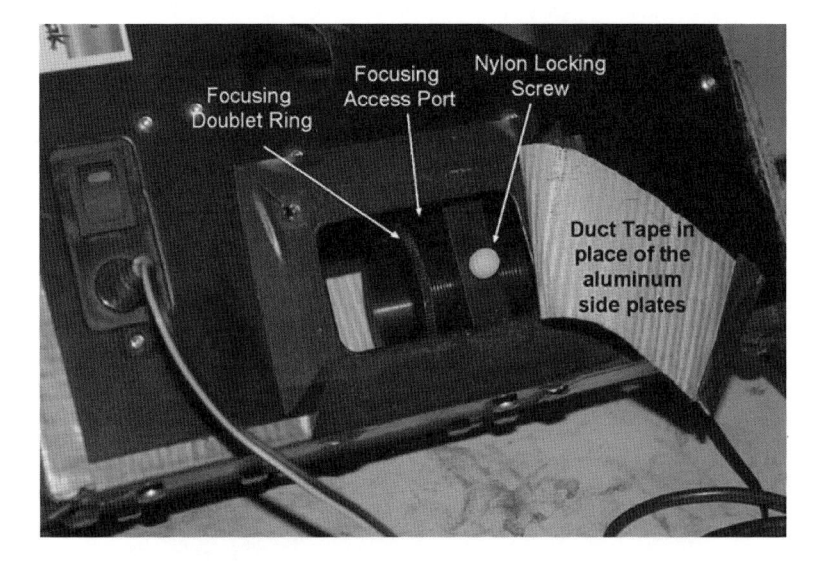

Fig. 6.10 Lhires III neon line focusing doublet

Adjusting the Spectrum Focus

Two things should be kept in mind when attempting to focus the spectrum. First, unless the focus is close, it is very possible even the neon calibration lines cannot be seen. If the focus is too far out of adjustment the spectrum will be spread out so far that nothing can be seen. If this is the case, try using a longer exposure until something is seen to focus on. Second, adjusting the focus will cause a wavelength shift. Never adjust the focus between taking the calibration spectrum and stellar spectrum. When using a 2,400 l/mm grating and going to a different area in the spectrum, say from hydrogen alpha to the hydrogen beta region, significant refocusing will be required. Even the 600 l/mm grating will require some refocusing when changing the wavelength (Fig. 6.11).

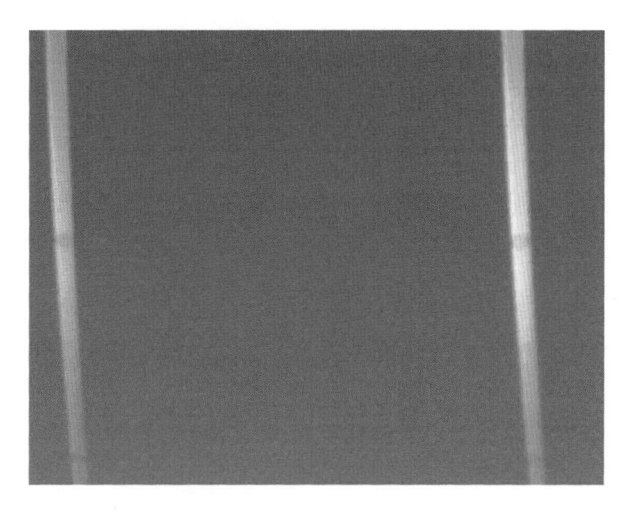

Fig. 6.11 Unfocused neon lines

There are three things that will be noticed with elemental calibration lines. First, the built-in neon calibrator line spectrum goes from top to bottom whereas a stellar spectrum is a thin horizontal line. This is normal because the neon spectrum fills the whole slit at one wavelength and the stellar spectrum fills just a small part of the slit over the whole spectrum. Next, it will be noticed that the neon lines are curved. This is normal and caused by the Littrow design. Because of this line curvature it is very important to uses the same delimiting area of the spectra for both the calibration lines and the stellar spectrum. When doing the calibration, first use the spectrum delimiting lines to just bracket the spectrum. Then without changing the horizontal lines positions, load the neon lines and use that same area for the calibration. If the horizontal lines are moved up or down, the wavelength calibration will have an error increasing the further from the stellar spectrum due to the neon line curvature.

The last thing to notice is that there is a notch midway in the calibration lines. This is where the slit is and is normal.

The neon lines should be focused to make them as narrow as possible. If the lines are bright, reduce the exposure time to better see the results. While the initial focusing can be done indoors, once the spectrograph is on the telescope and ready for spectroscopy the spectrum focusing should be checked again (Fig. 6.12).

Fig. 6.12 Focused neon lines

High-Resolution Imaging Technique

Lhires III Modifications and Adjustments

Bright star spectroscopy will produce a large star image. When the star image is on the slit, it will be too big to fall in. In fact it may be difficult to see any change and to know when the star is on the slit and when it has drifted off the slit. One technique that works is to adjust the exposure time of the guiding camera. Then use the software to show a histogram or other readout for the peak ADU counts in the image. Adjust the exposure with the star off the slit so the counts are slightly below 65,535. Now move the star onto the slit. The ADU counts should fall significantly. For example, when imaging Betelgeuse the exposure may be set to just a few milliseconds and the counts off the slit in the range of 50,000–60,000. When moved onto the slit, little change is seen in the star's image, but the histogram counts drop to between 10,000 and 20,000. This provides an objective measure that can be used to know when the star image has drifted out of the slit. Remember this is for the guide camera and doing this has no effect on the spectrum, other than to shorten the spectrum exposure by keeping the star in the slit. Another technique that helps is using a focal reducer. This will make the star image smaller and the light more concentrated. The whole purpose is to get as many photons from the star into the slit as can be done for the exposure time.

Star Image Focusing

This focusing is done with the telescope focus. Note, the star image as seen in the guide camera will not be high quality. This is due to the use of a polished metal reflective slit. Move the star image close to the slit and use the telescope focus to focus the star image as sharply as possible. The image may be far from symmetrical, but the trick is to make the image as small as possible. Note if the focus is very far above or below the slit it will be different from the focus of the star near or on the slit. When taking the neon calibration image and using the external neon ring, be sure to move the star slightly above or below the slit while taking the calibration image. Once the neon calibration has been imaged do not touch the telescope or spectrograph. Even a slight stress can cause a flexure that will change the calibration. Use the telescope's electronic remote control. This is why using the external neon ring is good.

Spectral Order

With the 2,400 l/mm grating the spectrograph is set to view only the first order spectrum. With other gratings it is possible to view higher orders above the first order. While they will work, they will be considerably dimmer than the first order spectrum. Always use the first order spectrum.

For the 2,400 l/mm grating the hydrogen alpha line at 6563 Å is centered in the image with the micrometer set at about 2,000. With the 600 l/mm grating the hydrogen alpha line centered at around 990. For other gratings one should start with the micrometer set close to zero and move out until the first order spectrum is seen.

Additional Spectrum Processing Considerations

There are three more considerations for processing a high-resolution spectrum image, image rotation, background subtraction and binning. For more information on these see sect. 2.3.

High-Resolution Spectrum Processing

Wavelength Calibration

To be of scientific value the line profile must be wavelength calibrated. In principal this is very simple, just associate each pixel X-axis (column) position with a wavelength. There are several ways to do this. The requirements for medium and high-resolution calibration are a bit different from the low-resolution calibration. First, the zero order spectrum will not be seen, second, because only a small window of the spectrum is imaged the linearity is fairly good so a linear calibration with two or more points works well. If more than two calibration lines are available it is still best to do a non-linear higher order fit, however. Usually using the dispersion and one point does not produce good results for high-resolution calibration. If neon lines are not in the window or not enough of the lines, other elements can be used, but add complexity. Sometimes telluric lines (atmospheric water vapor) lines can be used for the calibration. Note, the neon lamps typically have a fair amount of argon gas in them in addition to the neon gas. If the lines are identified they can be used as a reliable calibration line.

Selecting the Calibration Area

As mentioned earlier, the neon lines will appear curved. This is normal. The stellar spectrum will be seen as a horizontal image on the computer screen. The vertical position of the spectrum on the screen is determined by where in the slit the star was when the spectrum image was taken. Moving the star position right or left in the slit will cause the resulting spectrum to move up or down in the display. The position of the stellar spectrum is not critical, but should be below the top and above the bottom of the screen and not in the area where the notch in the neon lines is. Once the calibration and stellar spectra have been taken, when processing the spectrum, it is very important that the same horizontal area be used for the wavelength calibration. This is because of the curved neon lines. Determine the area of the spectrum image for the line profile first for the stellar spectrum. Do not change those orange horizontal delimiting lines when loading the neon spectrum. This will assure a more accurate wavelength calibration (Fig. 6.13).

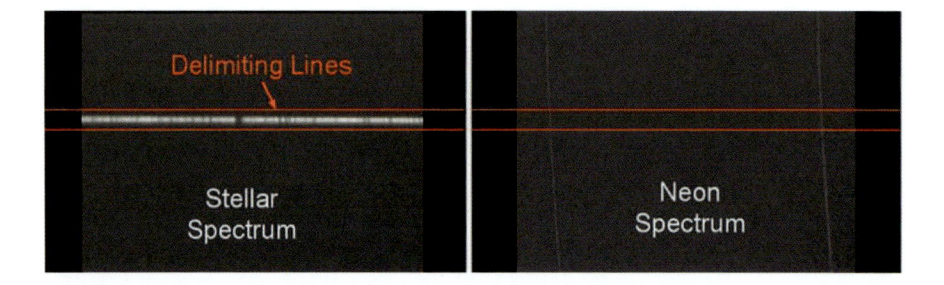

Fig. 6.13 Spectrum delimiting lines

Neon Calibration

The Lhires III built-in Neon lamp calibrator is very handy for getting familiar with the instrument and for focusing the spectrum. Its main purpose is for calibrating the wavelength of the spectrum line profile. One problem with using the internal neon calibrator for high-precision calibration is that it requires the user to touch the spectrograph to switch the flip mirror to and from the calibrate position. While some people suggest take the calibration before and after the main spectrum imaging, this may still not be as accurate as can be. Even very light touching of the spectrograph or motion of the telescope other than tracking motion can shift the calibration (Table 6.2).

High-Resolution Spectrum Processing

Table 6.2 Neon line pixel x-axis position with internal neon calibrator

Star #	Ne 6532.88 Å	Ne 6598.95 Å
1	179.34	781.65
2	179.77	780.99
3	177.70	779.27
4	180.94	782.50

Neon Ring Calibrator

A more accurate means of using a neon lamp calibration has been devised at the Hopkins Phoenix Observatory. A ring of eight neon lamps mounted on a wooden ring can be constructed. The neon lamps with current limiting resistors can be purchased three for $1.00 from All Electronics, http://www.allelectronics.com/. While plastic or metal can be used, metal is not recommended due to possible shock hazard if any bare wiring should touch it or if there is electrical leakage. An insulating material, such as wood or plastic, is best. The neon lamps can have built-in resistors allowing them to operate directly off of 110 VAC. For 220 VAC use sets of two neon lamps, with resistors, wired in series and the sets in parallel. The line frequency is unimportant and 50 or 60 Hz is fine. If more intensity is needed, more neon lamps can be used. All connections should be soldered and heat shrink tubing and or electrical tape used to completely cover any exposed wires (Fig. 6.14).

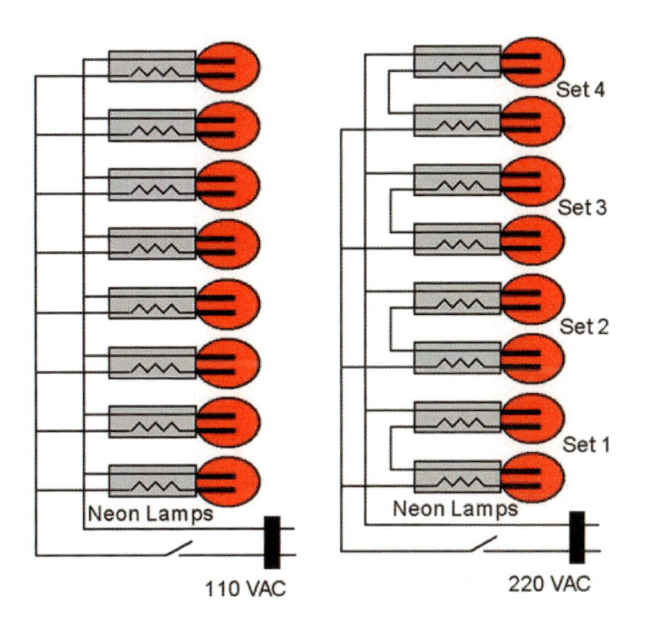

Fig. 6.14 Neon lamp ring wiring

The ring can be placed on the front of a telescope. The lamps can be turned on and off remotely without touching the telescope or spectrograph. There is also no need to turn the calibration flip mirror on the Lhires III. It was found a 30 second exposure on a 12″ telescope with the eight neon lamps produces a good neon calibration spectrum (Fig. 6.15).

Fig. 6.15 Telescope neon lamp ring calibrator

An experiment was done using this external neon ring and a 2,400 l/mm grating centered on the hydrogen alpha region. At least two good neon lines bracket the hydrogen alpha line. These were used to see if the neon line pixel positions changed when moving the telescope from star to star and doing the calibration. It was found that the pixel positions did not change with a before and after calibration if the telescope and spectrograph were not touched, but did change significantly when moving to a new star. The following table shows the neon calibration when using the internal neon calibrator. The numbers under the wavelengths are pixel positions of the neon lines in the line profile.

The measurements were done using the Measure Lines of RSpec to find the barycenter pixel X-position of the line. While the positioning of the Measure Lines is somewhat subjective, the data were very closely repeatable. Details on using the measure lines will be discussed in Chap. 7.

Low-Pressure Sodium Light Calibration

To be able to calibrate the sodium D line region precisely, a low-pressure sodium light can be used. By searching the Internet one system was found that seemed nice, but had an outrageous price tag, $450. The low-pressure sodium lamp, an 18-W Low-Pressure Sodium Lamp, Philips 2 34047 SOX- E18, is not too expensive at under $50 (Fig. 6.16).

High-Resolution Spectrum Processing 167

Fig. 6.16 18-W low pressure sodium

Initial searches for a power supply turned up little. Most ballast units were over $200 without the lamp or socket. Prolight has one for $150. The tube, tube socket and ballast were ordered. The light arrived broken. Another light was ordered through Amazon.com for $30. An online note was discovered about a digital ballast/power supply for $20 from Fulham, www.fulham.com/. An order was placed for both the ballast and another lamp. The replacement lamp came first.

The replacement lamp and socket were wired to the Prolight ballast. The Prolight ballast has a large transformer and large capacitor. A wooden case was built for it. The Prolight ballast, while expensive and bulky, works fine.

Fulham Digital Ballast Wiring

When the Fulham digital ballast and tube arrived they were wired together. Surprisingly, the digital ballast works as well as the larger much more expensive Prolight ballast. Two bright low-pressure sodium lamps are now available for calibrating the sodium D lines at the Hopkins Phoenix Observatory (Figs. 6.17 and 6.18).

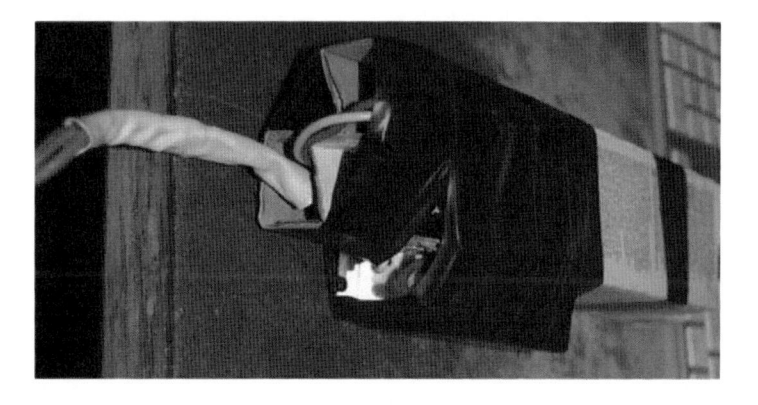

Fig. 6.17 Fulham digital ballast

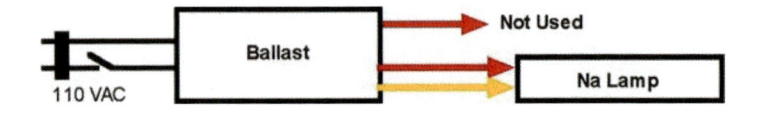

Fig. 6.18 Fulham digital ballast wiring

The black and white wires are for the 110 VAC input. The yellow and one red wire are connected to the tube. The second red wire is not used and should be capped and insulated. Since the digital ballast is small and light, a box was not made for it. To protect the tubes the cardboard container that the tubes came in was used. The digital ballast was taped to the container with black electrical tape. A window was cut in the container for the light. Lots of black electrical tape was used to secure and insulate all wires and the ballast. A switch can be used, but it is simpler just to not have a switch and just plug the unit in when desired. The tube is just plugged in or operated by a remote switch. It should be noted it can take 15 min or more for the tube to warm up. If the tube has not warmed up properly the spectrum will not be correct. Holding it in front of the telescope for a 10-second exposure works well (Fig. 6.19).

Fig. 6.19 Digital ballast wiring and glowing sodium light

Lhires III Tips

The best way to learn about the Lhires III is to experiment with it on the bench. Use the built-in neon calibrator to create the neon spectrum. Experiment with the focus using the focusing doublet accessed through the sides of the spectrograph. Orient the imaging camera properly. Do a wavelength calibration of the micrometer, micrometer setting versus wavelength for each grating used with the Lhires III. Plot a graph. While this will not be a precise calibration, it can get you close. Use red, green and violet laser pointers to help. Remember that going from one area of the spectrum to another will require significant refocusing. Remember also that the neon lamps have other gases in them, namely argon so some lines seen may not be neon lines.

Lhires III Tips

Do not plan to use the Lhires III on a telescope until you are comfortable with it on the bench. Trying to get familiar with the spectrograph at the telescope at night can be an extremely frustrating activity.

Once you are familiar with the Lhires III it is time to give a try on the telescope. While most telescopes can be used anything less than an 8″ telescope may not be able to support the Lhires III with two CCD cameras. Ideally the telescope will have a permanent mount and already be properly aligned. An enclose observatory is a welcome addition and will make the observing both easier and more pleasurable. If you cannot do that, just a pipe in the ground with concrete and a metal top for a wedge and telescope can help greatly.

On the bench and before the observing session, set the Lhires III micrometer to the approximate area of the spectrum of interest. This can be adjusted further at the telescope once a known spectrum is seen.

When at the telescope the first thing to do is turn on the CCD cameras and computer(s) to let them stabilize. Make sure the flip mirror is in the star position. Next find a bright object, a planet or even the moon. Find the object in the guiding camera and focus it with the telescope focus. If the focus is too far out of adjustment, even the Moon can be a challenge.

Next set the imaging camera for a 1.0 second looping/continuous exposures. Put the bright object on the slit about midway between ends of the slit. This is a critical point. You should see a faint spectrum. If no spectrum is seen you must find out why not. Make sure you are seeing the complete image on the computer screen. Sometimes the spectrum is there, but the image must be scrolled up or down on the screen to be seen. If that isn't the problem check to see if the flip mirror on the Lhires III is in the star position. If still no spectrum, put the flip mirror in the neon calibrate position and turn on the built-in neon lamp. Since a neon spectrum could be seen on the bench, it should work at the telescope. This gives you a chance to check where the problem lies. Perhaps the telescope is covered. Once you are seeing an astronomical spectrum you are almost there. This will be a good time to fine-tune the adjustment of the orientation of the imaging camera to produce a horizontal spectrum. Once set, the cameras should not need future adjustments.

Now look at a bright A type star, e.g., Vega. If the micrometer has been set to see the hydrogen alpha line a large absorption line should be seen. Make adjustments as needed. If the telescope provides enough photons and the star is bright enough, you can still use the 1.0 second looping exposure while you adjust the wavelength with the Lhires III micrometer. If the image is too faint, increase the exposure time. An increased exposure time will make the spectrum brighter, but slow the adjustments. Once you have a good spectrum in the desired wavelength region you are ready for your first serious spectrum image. Use the techniques discussed earlier to determine the proper exposure time for the spectrum.

Chapter 7

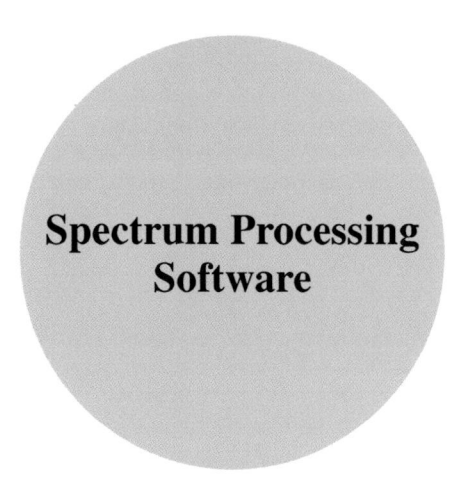

Spectrum Processing Software

Introduction

There are two kinds of image processing software used for spectroscopy. The first is a camera control and imaging processing program. Most CCD cameras come with software that can do this. The Meade DSI cameras come with AutoStar Suite software. The Suite has several parts, a planetarium program, telescope control, CCD camera control and image processing. The Orion StarShoot G3 cameras come with a suite like the Meade's. It is called the Orion Camera Studio. This program controls the Orion CCD camera, acquires and processes the image and can do further image processing. A commercial program is available called MaxIm DL (http://www.cyanogen.com/). The cost varies from $200 to $665. Unless you have specific need for a program like MaxIm DL, the software that comes with the CCD camera is probably sufficient for the acquisition and image processing.

Once you have a spectrum image and have the initial calibration done, such as dark frame subtraction, a second program is needed to process the spectrum. What is interesting is the spreadsheet program Excel can be used to do DIY (Do-It-Yourself) spectrum processing. While this works, it adds a large degree of complexity and added work. It may still be of interest to see what is going on and better understand the other spectrum processing programs.

The packaged spectrum processing programs may be limited to the spectrum processing, meaning they may not provide drivers for cameras and do not do image processing, such as dark subtraction. The programs discussed here are first the DIY approach using Excel. Next, is a commercial program called RSpec and then a French freeware programs called VSpec.

Data Reduction Versus Spectrum Processing

There has been some discussion as to what is the correct way to call the production of a calibrated line profile from a spectrum image. Some say it is data reduction and others say it is spectrum processing. Which is politically correct? There is probably no correct answer and it all depends on the political part. It would seem that logic dictates the correct answer is spectrum processing, but there will surely be those who resist. So where do these terms fit in?

Data Reduction

Before spectroscopy became popular with amateur astronomers, astronomical photoelectric photometry was the main way for an amateur astronomer to make valuable astronomical scientific contributions.

The goal of astronomical photometry is to determine the extraterrestrial magnitude (the corrected magnitude of the star if it were measured outside the Earth's atmosphere) of the star. A star's brightness is usually measured in specific bands (UBVRIJH). This measurement could be done using a single-channel photometer, such as a PMT photon counter, voltage or current averaging PMT or a PIN diode. The output of these devices is an electrical signal that is converted to a number. This number can be a photon count per second, a voltage or current average per second or a chart recorder amplitude reading per second. For multi-channel photometry, such as CCD photometry, a number is produced that represents the sum pixel ADU counts of a star's image. These numbers are first corrected by subtracting a similar number representing the sky intensity around the star. The resulting net number can be called "I" for representing the intensity of the star.

There are two different magnitudes involved in determining the extraterrestrial magnitude. They are the raw and instrumental magnitudes. The following is for the V or visual band, but other bands are similar. The raw magnitude of the star is calculated by:

$$\text{Raw Magnitude}(\text{RM}) = -2.5\log(\text{I})$$

Assume I = 35,000

$$\text{RM} = -11.360$$

Next an instrumental magnitude "IM" is calculated to correct for the instrument's sensitivity. The following equation determines the instrumental magnitude.

$$\text{Instrumental Magnitude}(\text{IM}) = \text{RM} + \text{Zp} + \text{epsilon}(\text{B}-\text{V})$$

A 16" telescope measuring the same star and at the same time from the same location as a 6" telescope will produce a much brighter raw magnitude. A correction for

Spectrum Processing

this difference must be made using a system zero point. The zero point represents the system's sensitivity. The zero point (Zp) for the system must be determined and added to the equation.

For this illustration assume Zp = 15.654

Next a color transformation coefficient must be determined. The filter used is the main contribution to this color coefficient, but other parts of the system may contribute to this also. For the V band the color transformation coefficient is called epsilon. For this illustration assume epsilon = 0.015. Epsilon is then multiplied by the (B-V) value (previously determined) of the star. Assume (B-V) = 0.235.

$$IM = -11.360 + 15.654 + (0.015 \times 0.235) = 4.297$$

Finally a calculation is done to correct for the starlight's attenuation (extinction) due to the Earth's atmosphere and produce the final extraterrestrial magnitude. This is done using the air mass X of the observation and a nightly determined extinction coefficient k'. The value of X is 1.00 at the zenith and increases somewhat exponentially as the direction approaches the horizon. Sometimes a second order extinction coefficient k' is used to increase the accuracy.

$$\text{Extraterrestrial Magnitude}(EM) = 4.297 - X \times k'$$

Where the following are assumed:

$X = 1.432$
$k' = 0.141$

$$EM(M) = 4.298 - 1.432 \times 0.141 = 4.095$$

The final magnitude "M" for the star is 4.095. This has been "reduced" from the initial Raw Magnitude Intensity of 35,000.

Spectrum Processing

When a spectrum image is taken it must undergo two stages of processing. If multiple images are taken they must be stacked. Then if the exposure is longer than 1.0 s dark frames must be taken and subtracted. If flat frames are used they must be divided into the image. This is regular image processing not image reduction. The resulting image can now undergo spectrum processing. Because the spectrum is just a small part of the whole image the first step is to delimit the area of the spectrum. Next, a subtraction of the sky immediately above and below the spectrum is done. If so desired, a horizontal binning is next done. The delimited image now has the pixel column ADU counts summed. This then produces an un-calibrated line profile. The intensity (in ADU counts) is the vertical Y-axis and pixel column position is the horizontal X-axis. The line profile can then be wavelength-calibrated. This is indeed a processing technique as opposed to a data reduction technique. There is no initial number that is reduced to a final number.

DIY Spectrum Processing (Excel)

Introduction

If a spectrum line profile is created with the DIY spectrograph it can be saved as a text file. There is a small header with the scan date (this uses the computer's date and time), Integration Time (exposure time) and Number of Averages. Below the header there are five columns of data. There will be 2,047 rows of data corresponding to the 2,047 pixels on the linear CCD chip. In the Table below the profile has not been calibrated so the wavelength (λ nm) is the same as the pixel #. Note, for the DIY Spectrograph the wavelengths are in nm not Å (Fig. 7.1).

Header Data				
Scan Date	09/29/2012 15:31			
Integration Time	150			
Number of Averages	1			

Spectrum Data				
Pixel #	λ nm	Sum	Avg	Background
0	0	735	735	0
1	1	732	732	0
2	2	743	743	0
3	3	742	742	0
.	.	.	.	.
.	.	.	.	.
.	.	.	.	.
2047	2047	754	754	0

0	735
1	732
2	743
3	742
.	.
.	.
2047	754

Fig. 7.1 DIY data text file and trimmed file

The data text file can be loaded into Excel. Only two of the columns are needed, Pixel # and Sum. To create a line profile in Excel, select the two columns and choose the graph desired. For data files with multiple columns of ADU counts, Excel can be programmed to produce another column which is the sum of those counts for a given pixel position. That sum and pixel position can be used to create the line profile graph. The pixel # versus wavelength can be determined and a new column created that shows the wavelength for each pixel position. The Sum column and new wavelength column can then be used to create a wavelength calibrated line profile graph (Fig. 7.2).

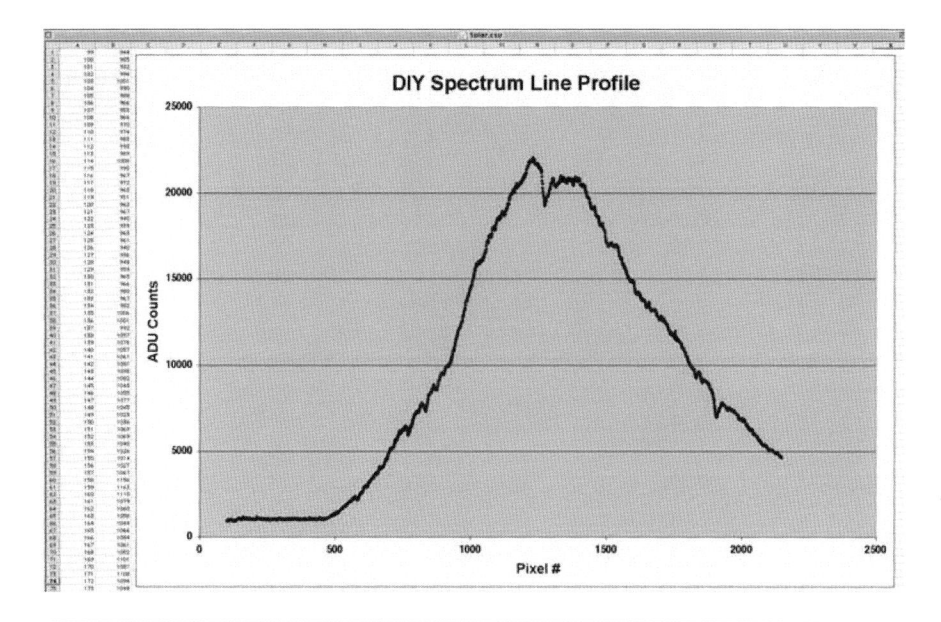

Fig. 7.2 Excel line profile of DIY spectrum data

As noted earlier, it is easier to use RSpec or VSpec to process the spectrum, but using Excel can help understand what is going on.

To open the DIY trimmed data file in a program, such as RSpec or VSpec, first delete the header data, column titles and all columns except the Pixel # and Sum. Save the resulting file as a text file. Now change the extension from .txt to .dat. The file can now be opened in RSpec or VSpec. Remember this is no longer an image file so it must be opened as a line profile .dat file. Once opened in RSpec or VSpec the profile can be wavelength calibrated and further processed.

RSpec Spectrum Processing Software

Introduction

RSpec is the only commercial spectrum processing software package available at this time http://www.RSpec-astro.com. The latest version of the program, as of February 2013, is 1.7.0. One may ask why he or she should pay for a program when other similar programs are available free. It's like the old adage, "You get what you pay for." RSpec is a refined commercial program. Is it perfect? No software is perfect. The difference between RSpec and the freeware programs is that instead of spending many frustrating hours or days trying to figure out how to do things with the freeware, within a few minutes you are up and running with RSpec. RSpec encourages experimentation and exploration without fear of crashing the program or getting totally lost. A large part of learning something like spectroscopy is the ability to experiment. The freeware programs tend to be very strict. While they are powerful, they are shy on the human interface. If you step off the narrow path you enter a mine field and usually crash the program or end up not knowing where you are or what you are doing. To complicate things, many times you do not know where the path is. Most of the freeware is French converted to English. To many people, myself included, the freeware program icons make absolutely no sense. The thought process required to use the freeware programs is not what I would consider intuitive. On the other hand, RSpec is a joy to use and encourages the beginner.

RSpec stands for **R**eal-time **Spec**troscopy. It was originally used for low-resolution work with a Star Analyser and a video camera. Getting real-time spectra is fun and can be inspirational. For something like meteor spectroscopy it is essential. While RSpec certainly works for video spectroscopy it also has excellent potential for more serious spectroscopy including high-resolution work.

The program works with any DirectX video camera. The program allows control and acquisition of video camera images. Because most Monochrome and Color CCD cameras do not use DirectX, they cannot be controlled from RSpec. It is actually better to use a program that comes with the camera to control the image acquisition and do initial image processing. While the video cameras can be of interest for most work it is better to capture spectra with still cameras, preferably 16-bit monochrome CCD cameras. The monochrome CMOS cameras work, but most observers prefer the CCD type. The images should be saved as .fits (not integer .fits or other image type files). Note: Integer fits have half the dynamic range of regular .fits images. This means if you exceed a pixel count of 32,765, the number turns negative. While RSpec can open other formats, the .fits file is specifically designed for astronomical work and contains more information than available with other formats. The use of .jpg or .bmp files not only do not include header information about the image, they are limited to 8 bits or 256 levels as opposed to the .fits 16 bit 65,536 levels.

Some people feel RSpec is a toy program. Just like the Macintosh computers that people thought were cute and not for serious work, RSpec's ease of use can be misleading. Indeed, it is easy and fun to use, but it is much more than a toy program. Some serious features have been added for high-resolution work, e.g., the ability to normalize line profiles and then measure equivalent widths. Hopefully future versions will include a means for calculating heliocentric correction and associated data as well the ability to view and edit FITS headers. These features will help elevate the program to a more professional level. Still for the beginner and even the experienced spectroscopist, RSpec is an excellent and powerful program. While RSpec is designed to be used with Windows, it works well on Macintosh computers with programs like Parallels running Windows XP.

RSpec can be downloaded free as a time-limited (30 days) full capability program as with most software, RSpec is continually being refined. The screen shots in this book are from the current version (1.7.0). There is no User Manual for RSpec. There are, however, numerous demonstration videos that guide the user through RSpec. More information on RSpec and the videos can be seen at: http://www.rspec-astro.com/

Using RSpec

You can start examining your spectrum immediately by just loading the spectrum image, bracketing it with the horizontal delimiting lines and watching a line profile of the spectrum image appear. However, it is best to set the program up for your specific needs before venturing further. Figure 7.3 shows what the user will see when opening RSpec and loading a spectrum image. The colors and line widths may be different and can be set to the user's preference. In the following Figure a sample spectrum has been loaded and displayed. When first opening RSpec only blank Spectrum and Profile Windows are displayed.

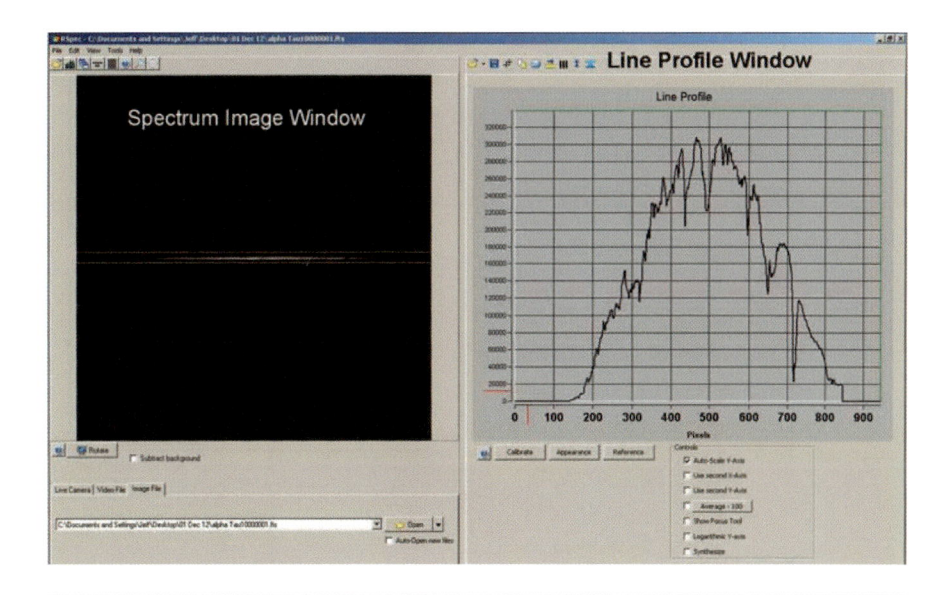

Fig. 7.3 RSpec spectrum image and line profile windows

Setting Options from the Spectrum Window

The Spectrum Window is the window to the left on the display. To begin there are some Options that should be set or at least checked. Things like the number of places for pixel positions and wavelengths should be set as well as some other options (Fig. 7.4). To do this, from the "Tools" tab at the top left, select "Options."

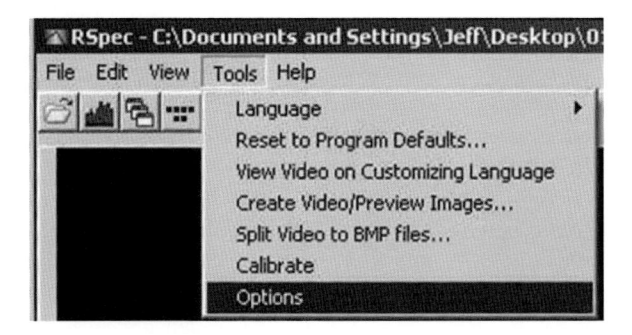

Fig. 7.4 Tools menu, option selection

The following selections in the Options window are suggested (Fig. 7.5).

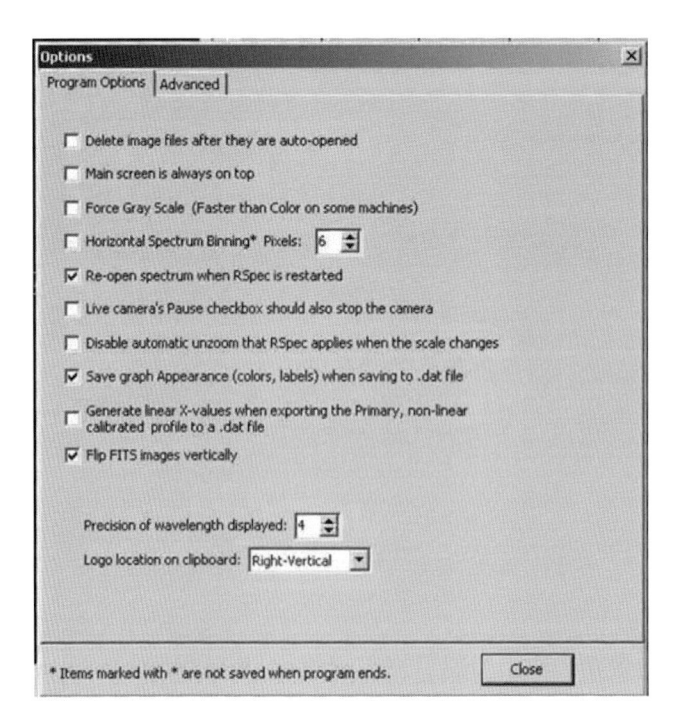

Fig. 7.5 Basic program options

There are a couple of options that need explaining. The "Horizontal Spectrum Binning" can be left unchecked and selected from the binning icon later. De-selecting binning will provide the most accurate display. For noisy images, binning can be used to smooth the line profile at the expense of wavelength resolution. The options are 2, 3, 4, 5 and 6. You can also select the binning from the top left menu bar.

The "Generate linear x-values when exporting the Primary, non-linear calibrated profile to a .dat file" is used when planning to export the .dat to a program like VSpec. Programs like VSpec have the limitation of requiring that the imported x-values must be evenly spaced. The non-linear calibration in RSpec will not produce equally spaced, linear, x-values unless this option is selected. Unless it is planned to export the data to another program, leave this unchecked.

The "Flip FITS image vertically" is not normally required as the spectrum image should be symmetrical around the horizontal axis. Some people have requested this feature, however. Typically it should make little or no difference if this is selected.

Set the "Precision of wavelength display" to 4. This will display wavelengths to 4 places, e.g., 6,562.3423 Å. While this is not important for low-resolution work, it is important for high-resolution work.

The rest of the options are obvious.

There are more options available by selecting the "Advanced" tab (Fig. 7.6).

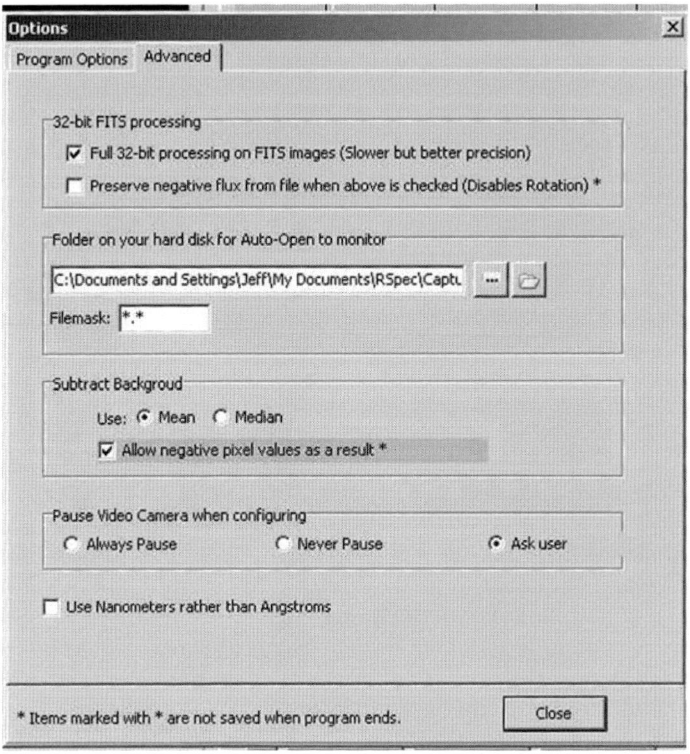

Fig. 7.6 Advanced program options

1. Select the "Full 32-bit processing on FITS images." Without this selected the processing will be faster, but with the newer computers there will be little or no difference seen.
2. For the "Subtract Background," use the "Mean" selection.
3. Do not select "Use Nanometers rather than Angstroms." For most serious work all wavelength values should be in Ångströms.
4. Select "Close."

Spectrum Window

The Spectrum Window is the window on the left side of RSpec. There are several more options in the Spectrum Window section that should be understood before proceeding.

Loading the Spectra

Now that you have a previously taken spectrum image you need to process it to produce a wavelength-calibrated line profile (Fig. 7.7).

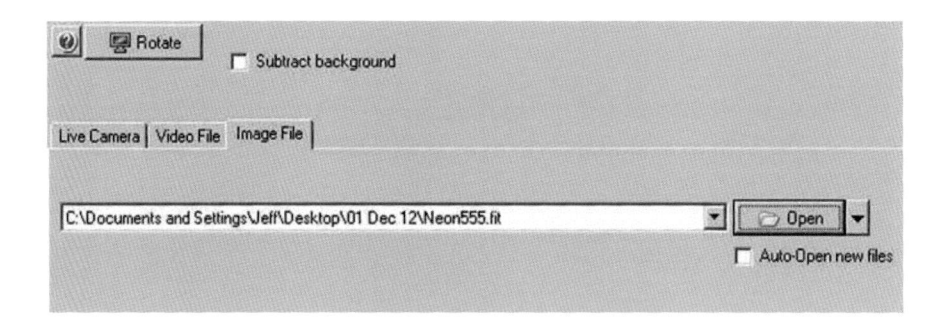

Fig. 7.7 File loading

The image source menu has three options for getting the spectrum loaded into RSpec. If you have a video camera connected you can get live images. A ToUcam works well, as do most DirectX video cameras. You can also load a video from a file or load a single image. For most cases a single image will have been taken and saved. For a single image select the Image File tab, find and load the image file.

For the following discussion a fixed image spectrum from the Image File will be used. If an exposure of more than 1.0 s was used the image should have a dark frame subtracted before loading into RSpec. These things must be done with the imaging software before getting to RSpec. RSpec does not have the capability to process images with dark frame subtraction or flat field calibration. RSpec does have an Averaging Option for multiple images that is similar to the stacking of the image processing programs. This can be selected from the Profile window Controls. The selections are 5, 10, 30, 50, 80, 100, 150 and 200 images with the 100 being default.

The subtract background is very important for most spectrum images, particularly the low-resolution images taken with a slitless spectrograph like the Star Analyser or with the ALPY 600 in slitless mode. This is because there is a great deal of background sky in the image. If you are doing high-resolution work and have calibration images of spectral lines from a neon calibrator or elemental gas discharge tube, you should unselect the Subtract background option. This is because the emission light fills the slit and will produce vertical lines from top to bottom of the spectrum window. Subtracting background will degrade or eliminate the calibration lines. Make sure the "Subtract background" is NOT selected when viewing the calibration lines. When viewing a stellar spectrum, the background should be subtracted.

Spectrum Delimiting Lines

The two horizontal orange lines in the Spectrum Window allow the spectrum to be bracketed/delimited. This is very important for low-resolution work with the Star Analyser as it is possible to have multiple spectra of different stars in the single image. Position the lines so that they bracket the spectrum as close as possible. It may take awhile to get used to moving the lines. Click on a line and hold the button to drag the line up or down. Clicking between the lines allows both lines to move up and down. A Shift-Click will also allow both lines to move. Whatever is between the orange lines will be converted to a line profile and displayed in the right Line Profile Window (Fig. 7.8).

Fig. 7.8 Spectrum window orange spectrum delimiting lines

Seeing the Spectrum

Select the "Image File" by clicking on the "Open" button. Find and load a spectrum image. In the following case a short exposure of Betelgeuse was used. This was a one second image of a high-resolution spectrum. For a 60 s exposure image or one where the spectrum can be seen the following procedure is not needed. With only a 1.0 s image the image below is what you see or rather don't see.

No Spectrum Seen

The spectrum image is too faint to see, but you can use the Histogram tool to adjust the view. This will not affect the line profile. This is very important for seeing faint stellar spectra. If you cannot see the spectrum you do not know where to put the orange parallel lines. By using the Histogram tool you can see the spectrum and position the lines (Fig. 7.9).

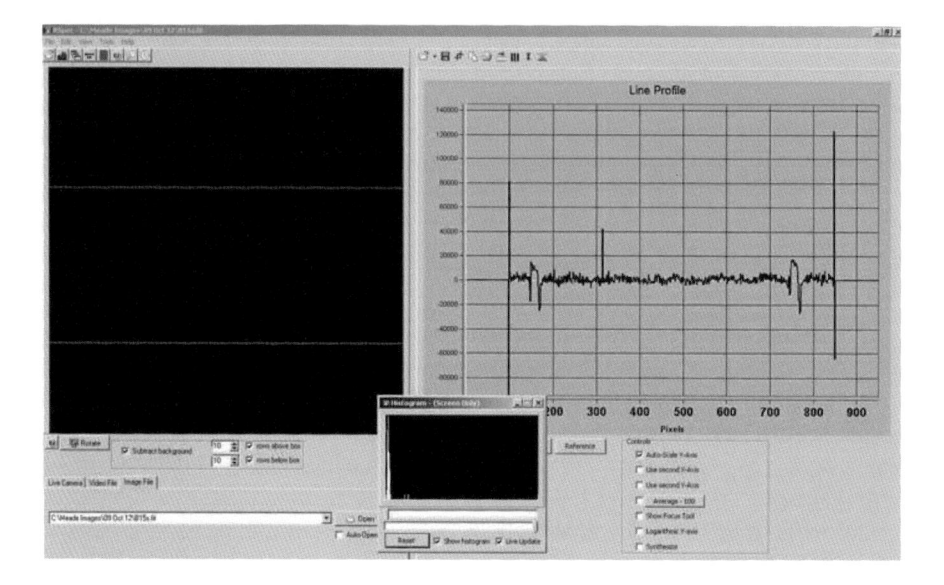

Fig. 7.9 No visible spectrum

The above line profile is just noise. If you move the orange lines up or down you may find a spectrum with sufficient SNR to produce a line profile. Another way to find the spectrum is to move one orange line all the way to the top and the other all the way to the bottom. That introduces the most noise along with the spectrum, however. Ideally the lines should bracket just the spectrum as close as possible to minimize the noise.

RSpec Spectrum Processing Software 183

Histogram

For help in seeing the faint spectrum at the upper left of the Spectrum Window select the "Histogram icon (Fig. 7.10).

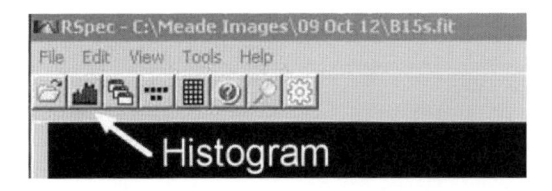

Fig. 7.10 Histogram tool

The Histogram window will appear. Adjust the sliders until you can see the image. Move the bottom slider to the left until the spectrum can be seen. This will cause the background to get bright too. Note, the adjustments to the image are not saved nor is the spectrum image modified. This also has no effect on the line profile. The sole purpose is to allow you to see the spectrum so you can bracket the spectrum closely with the orange horizontal delimiting lines (Figs. 7.11 and 7.12).

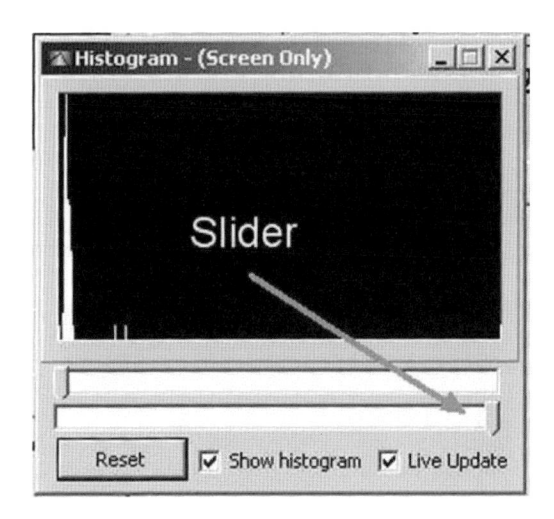

Fig. 7.11 Histogram sliders

184 7 Spectrum Processing Software

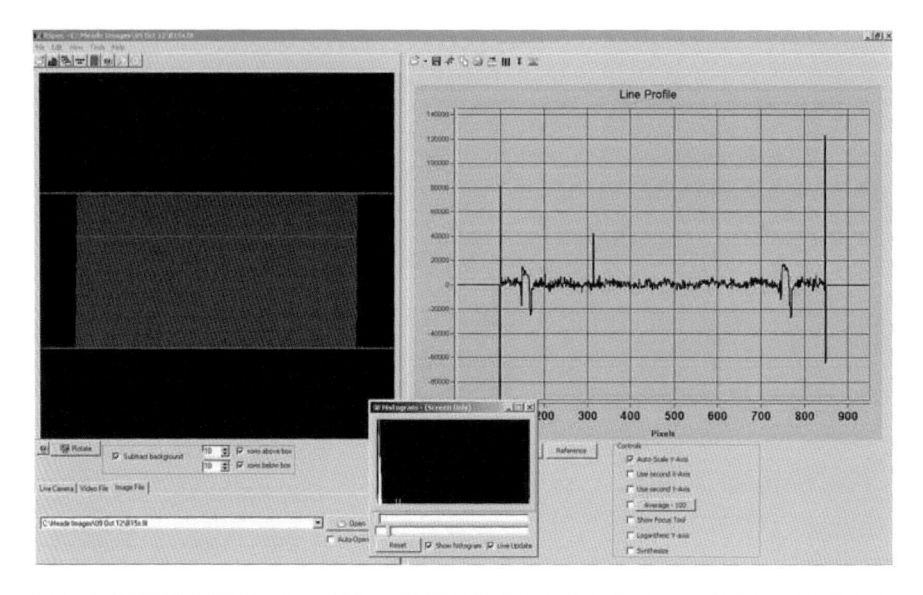

Fig. 7.12 Histogram image enhancement

Adjusting the Histogram slider to the left produces a brighter image and the faint spectrum can be seen. Although the above Figure shows a faint spectrum, because the horizontal orange delimiting lines are far apart, the line profile is still just noise, for high-resolution stellar spectra there will be a horizontal continuum. The trick is to use the parallel orange delimiting lines to closely bracket it and then subtract the background (Fig. 7.13).

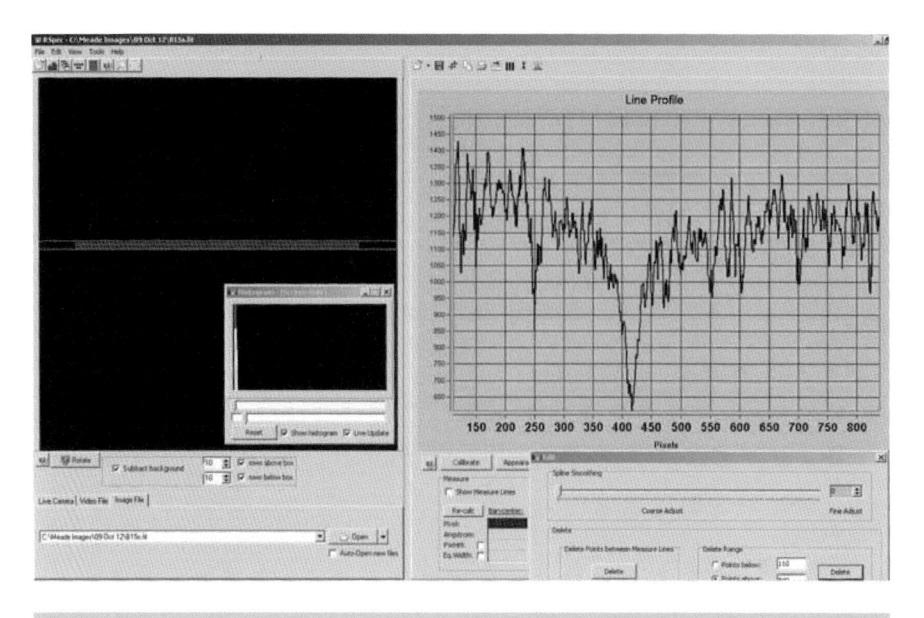

Fig. 7.13 Noisy line profile

The above Figure shows the spectrum image closely delimited by the horizontal lines. Whatever is between the lines will be shown in the right-hand Profile Window as a line profile. It may take some practice to drag the orange lines up and down, but the trick is to get them as close to the spectrum as you can. There is a "Zoom" option that can be selected from the small magnifying glass icon at the upper left of the Spectrum Window. The above Figure shows a noisy line profile of a hydrogen alpha absorption line. Using 6-pixel horizontal binning and trimming the end points allows a fair line profile to be seen. What is amazing even with the low exposure time and noisy profile, the line profile can be wavelength calibrated.

RSpec Pixel Map

RSpec version 1.7.0, released February 2013, has an added feature called Pixel Map. Once the spectrum image has been vertically delimited with the horizontal lines, using the Control and left button, a red box can be drawn on the spectrum to horizontally delimit the area for the pixel map. This reduces the number of pixels to just the area of interest. The map can be dragged larger to show more pixels (Fig. 7.14).

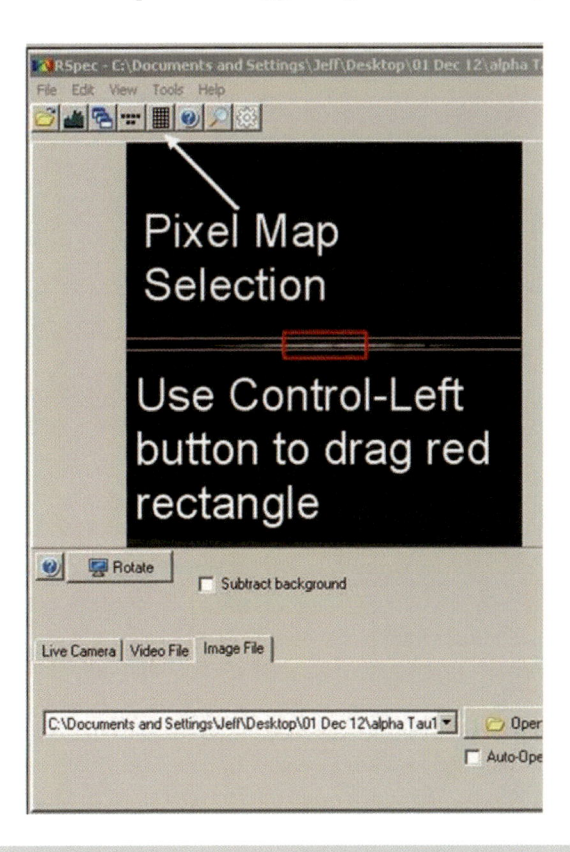

Fig. 7.14 RSpec pixel map selection

With 16 bit monochrome images the maximum ADU count will be 65,535. This is the point of saturation for the CCD or CMOS pixel. Exposure should be adjusted so that the maximum pixel ADU count is well below the 65,535 limit. For precise work, once the linearity break point has been determined, see section "Linearity", use that ADU count where linearity breaks as the upper limit and adjust the exposure to keep at or under that ADU count (Fig. 7.15).

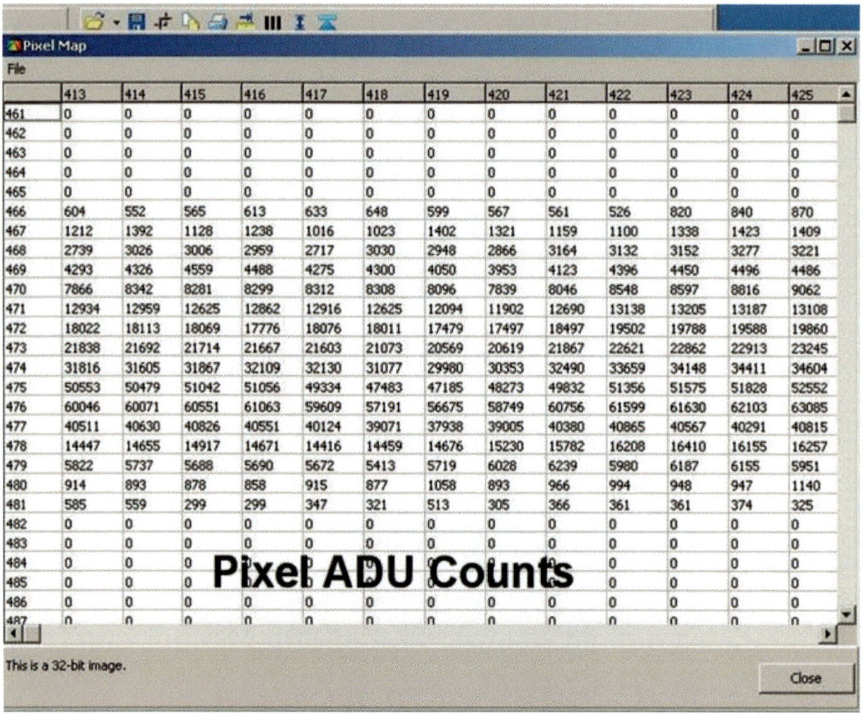

Fig. 7.15 RSpec pixel map ADU counts

Rotate and Slant of the Spectrum

For spectra that contain a continuum, e.g., from a star or other astronomical body, the spectrum will be a horizontal bar. It should be as close to horizontal as possible. This is best accomplished by adjusting/rotating the imaging CCD camera before taking the exposure. If for some reason this cannot be done, RSpec has 7.4 an option to allow you to rotate or tilt the spectrum until it is horizontal. For the case where you have emission lines, e.g., with the neon calibration, the lines should be vertical. If they are slanted RSpec provides a slant correction to get them vertical. With the Lhires III at high-resolution the neon lines will be curved. This is normal.

Sometimes when doing low-resolution slitless spectroscopy there may be background stars that interfere with the spectrum of interest. In such cases it may be necessary to rotate the image at the spectrograph to separate the desired spectrum from other stars. The software "Rotate" can then rotate the spectrum image back to the horizontal.

Note: When opening the program check to see if the Rotation is selected. Sometimes it is carried over when a new file is opened. If the image does not require rotation, set the sliders to zero.

To select the Rotate and Slant option, select the "Rotate" button (Fig. 7.16).

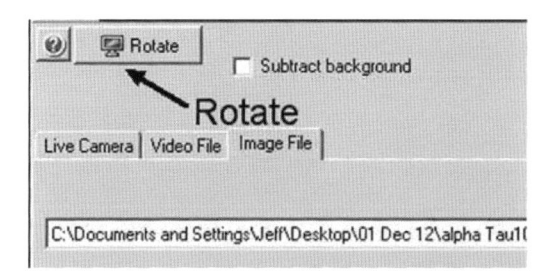

Fig. 7.16 Select rotate option

Note: Using the Rotate and Slant option can sometimes produce artifacts in the line profile so it is best to try to get the spectrum aligned by rotating the detector rather than using the software (Fig. 7.17).

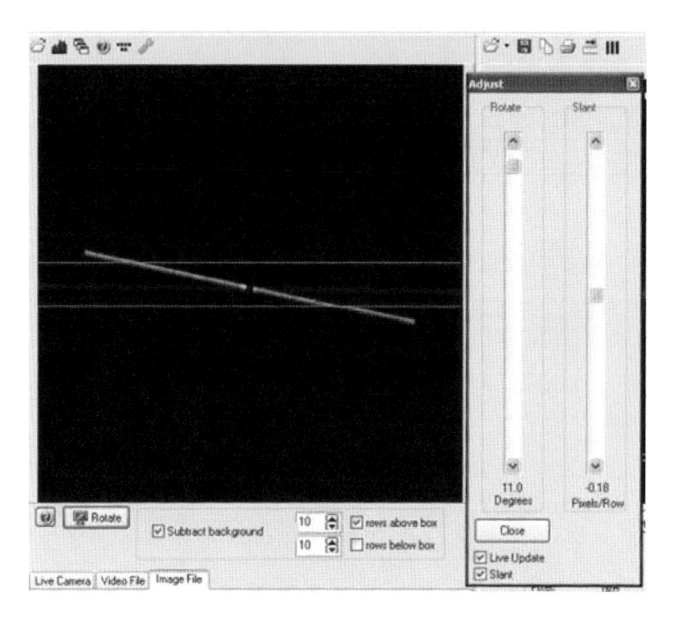

Fig. 7.17 Rotate and slant slides

Background/Sky

For slit-less spectroscopy, e.g., with a Star Analyser, there may be significant background (sky) light. In addition to bracketing the spectra as closely as possible, RSpec also allow subtraction of the sky around the spectrum. Even high-resolution slit-spectra benefit from the subtraction. Note, as mentioned earlier, when taking slit type spectra with elemental sources such as from a gas discharge tube or neon lamp, make sure the "Subtract background" is OFF, unselected. The spectral lines go from top to bottom because the slit is illuminated from one end to the other. Trying to subtract the sky just subtracts the whole line.

The following low-resolution spectrum image line profile shows the effect on a line profile without subtracting the sky (Fig. 7.18).

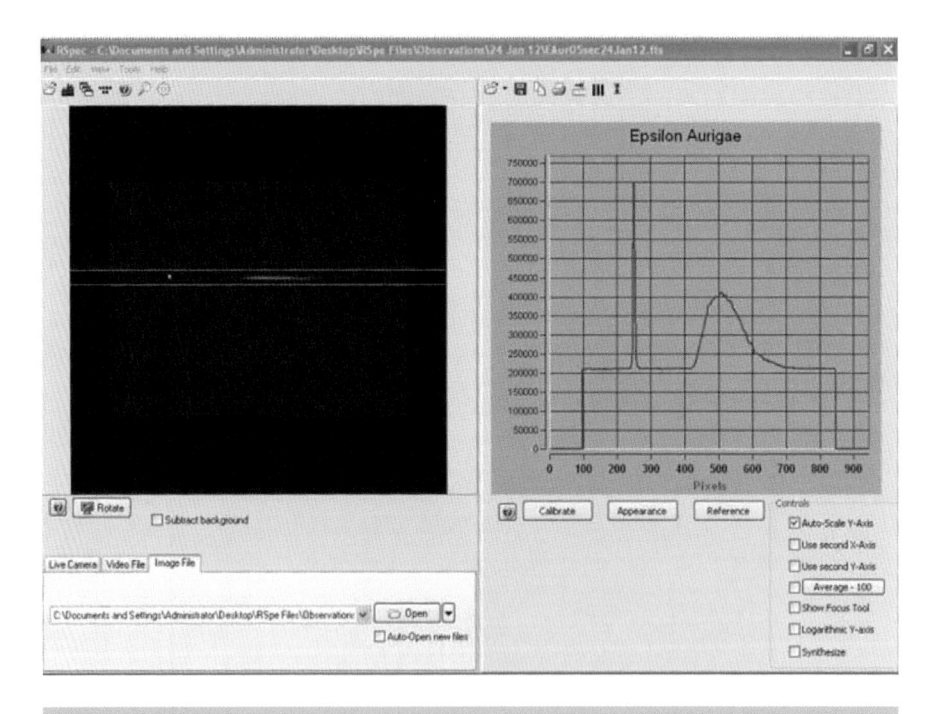

Fig. 7.18 No sky subtraction

Note the high floor level for the profile bracketing the zero and first order spectra due to the background noise.

Subtracting Background/Sky

The default 10 pixels above and below seem about right, but you can experiment with other values. The background subtraction is something to experiment with and see how it affects the line profile (Fig. 7.19).

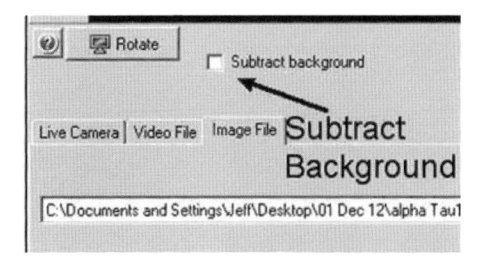

Fig. 7.19 Subtract background/sky option

The following image shows the effect of subtracting the sky (Fig. 7.20).

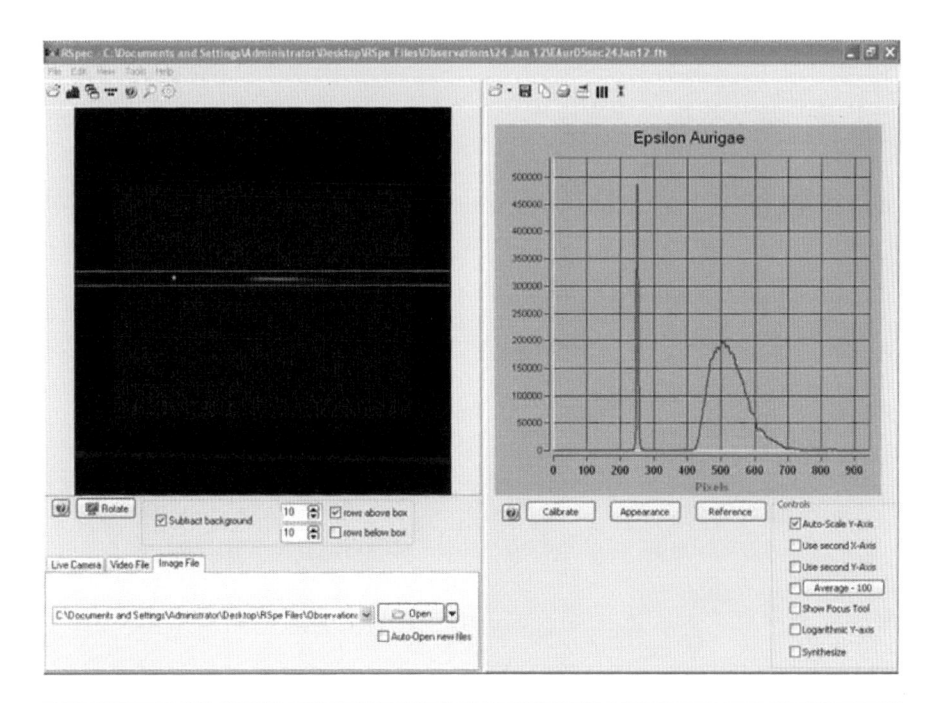

Fig. 7.20 Sky/background subtracted

Horizontal Binning

If the line profile is noisy, the use of different degrees of horizontal binning can smooth the profile. For low-resolution work, 6-pixel binning can help and will not significantly affect the wavelength precision. For high-resolution work the amount of binning should be experimented with and the lowest degree of binning that helps should be used. What horizontal binning does is to average adjacent pixels ADU intensity values (Fig. 7.21).

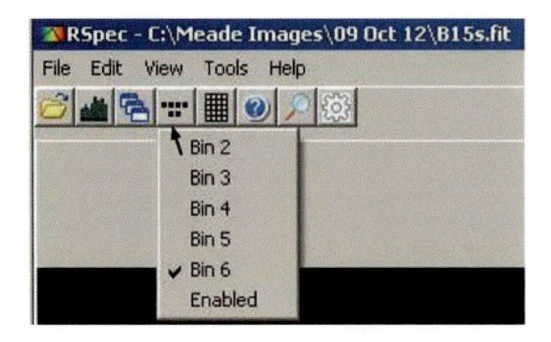

Fig. 7.21 Selecting horizontal binning

Line Profile Window

The Line Profile Window is the window on the right side of the screen. There are several options available from the line profile window that should be understood (Fig. 7.22).

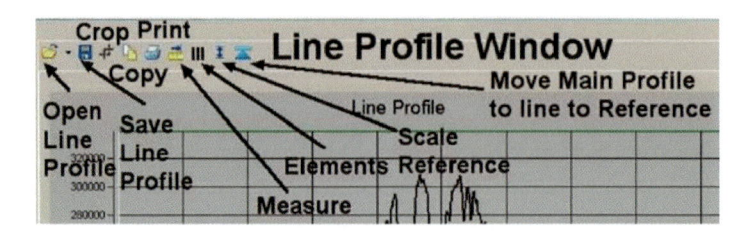

Fig. 7.22 Line profile window icons

RSpec Spectrum Processing Software   191

Note: The line profile window can be expanded horizontally by dragging the dividing frame edge to the left (Fig. 7.23).

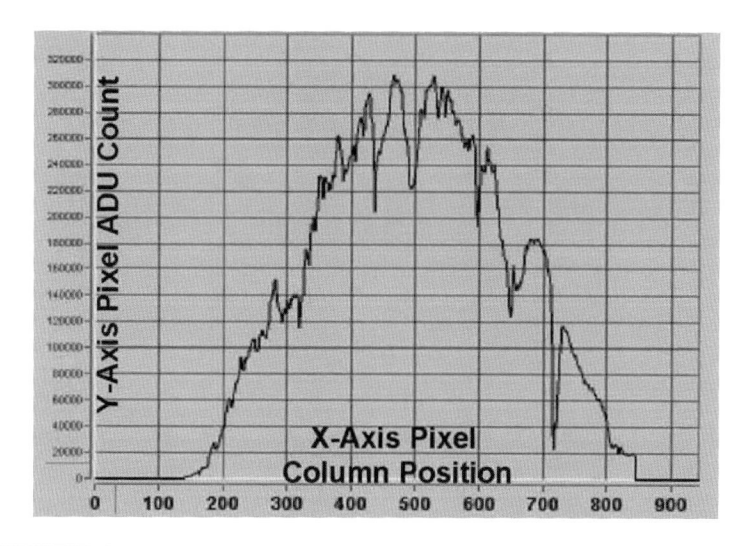

Fig. 7.23 Initial line profile window

Profile Versus Reference Files

RSpec uses two types of files for spectrum processing. The line profile file of the star or object of interest is called the **Profile** file and line profiles used for calibration are called **Reference** files.

Measure Menu

At the top left of the profile window there are several icons. The "Measure" icon when selected turns on a lower left display with items for making measurements of the line profile (Fig. 7.24).

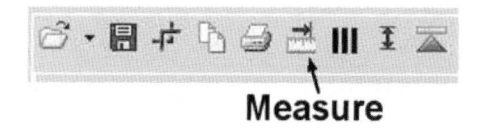

Fig. 7.24 Measure icon

Selecting the "Measure" icon displays the Measure information options (Fig. 7.25).

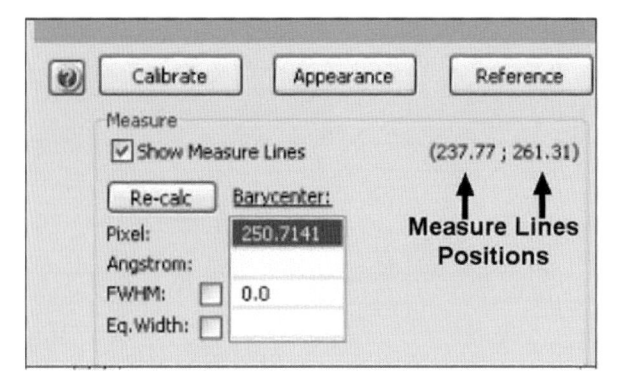

Fig. 7.25 Measurement information options

Measure Lines

One of the wonderful things about having a pixel mapped image is that you can determine a precise point between the pixel positions by determining what is known as barycenter. Barycenter is like a center of gravity. Selecting the "Show Measure Lines" allows two vertical delimiting lines to be moved to bracket an area to be measured. This is an important tool for processing spectra. The following Figure shows the Measure Line bracketing a hydrogen alpha absorption line (Fig. 7.26).

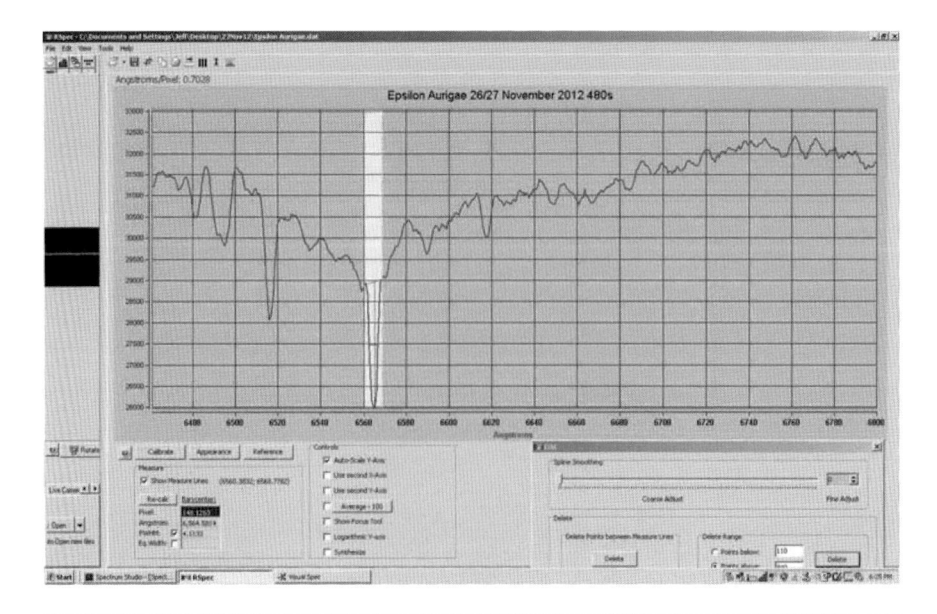

Fig. 7.26 Line profile window vertical measure lines

RSpec Spectrum Processing Software 193

Once the lines are set the measured results of the area between the Measure Lines is displayed. These are barycenter values (Fig. 7.27).

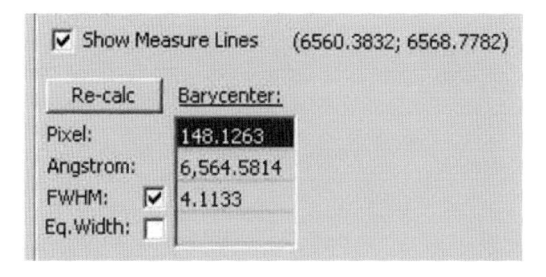

Fig. 7.27 Measure results

For the above profile the results show the center pixel x-axis position at 148.1263, the center wavelength is 6,564.5814 Å. If the FWHM is selected the value is 4.1133. Note, for the Equivalent Width (Eq.Width) the profile must first be normalized. This will be discussed later.

Using the Measure Lines

The Measure Lines can be used for several things. In addition to showing the barycenter of the pixel position, center wavelength and FWHM the lines can be used to delete or crop parts of the profile and at the ends to trim the profile. The Measure Lines are also useful when doing an instrument response calibration where specific lines need to be deleted to produce a smooth response curve.

Additional Selections

Just above the Measure Lines information are an icon and three buttons (Fig. 7.28).

Fig. 7.28 Additional selections

Appearance Menu

Selecting the left most icon shows a popup menu window with tips on zooming the image To enhance the appearance and annotate the line profile the "Appearance" button can be selected. Clicking on "Appearance" button allows "Labels," "Lines" or Graph options to be selected. RSpec has a wide range of features that allow you to control and modify the appearance and colors of the line profile. Most of the options can be left in the default state. Labeling the profile title from Graph option is good, however. When selecting the "Appearance" button the following windows are displayed (Fig. 7.29).

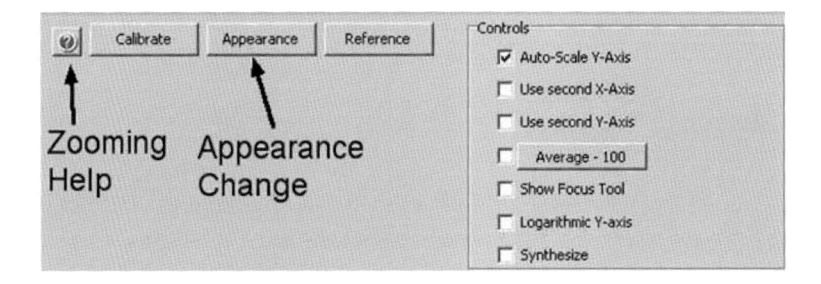

Fig. 7.29 Selecting appearance

Labels

The Labels window is first displayed. The Label window allows labels to be added to the line profile window. Labels such as line wavelengths can be useful and informative. This may require some experimentation to get the labels as desired (Fig. 7.30).

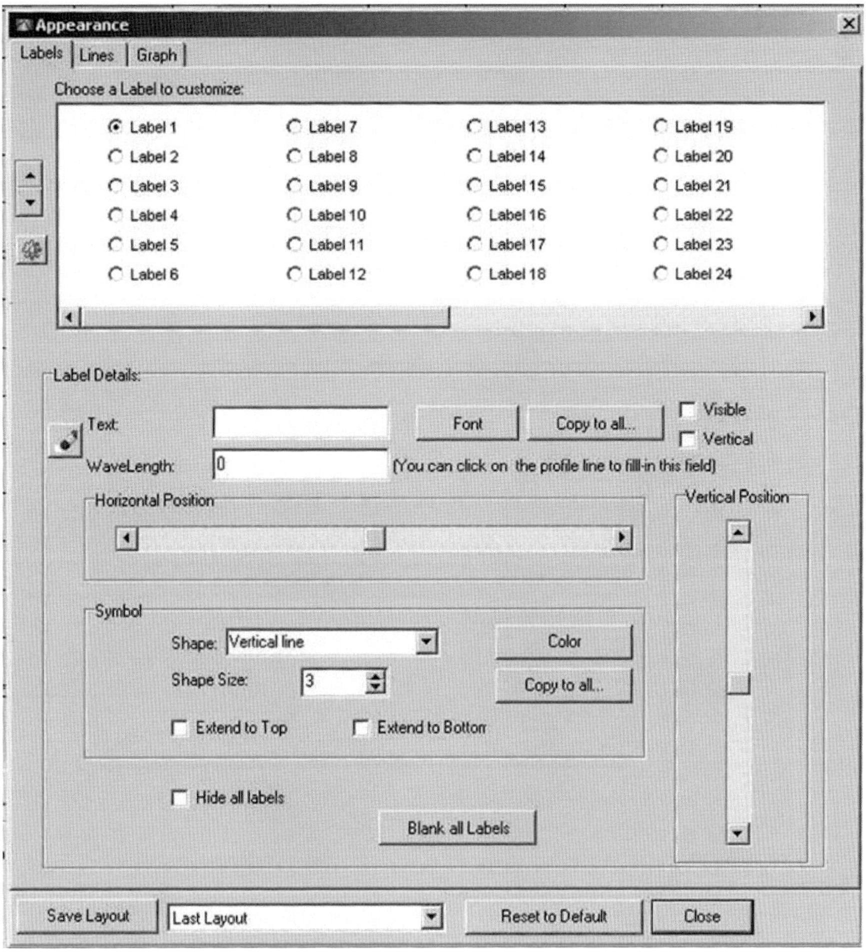

RSpec Spectrum Processing Software 195

Fig. 7.30 Labels menu

Lines

The middle tab at the upper left of the Appearance Window is the "Lines" tab. The Lines selection gives you the ability to control the width, color and point size of the lines in the line profile (Fig. 7.31).

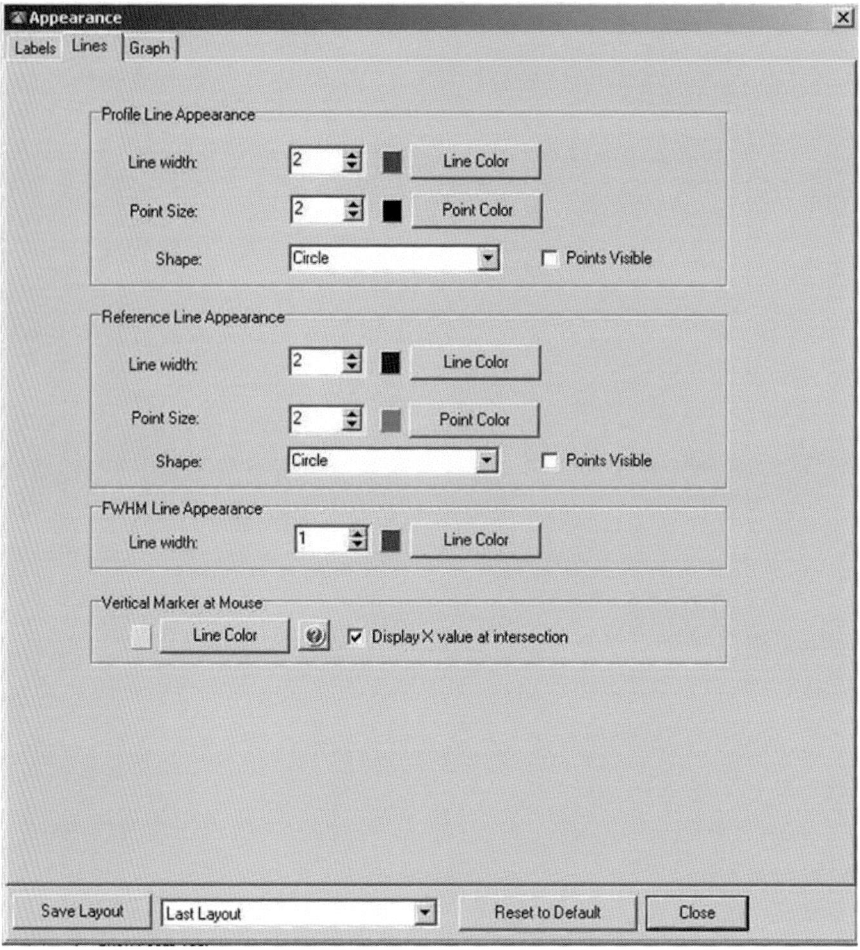

196　　　　　　　　　　　　　　　　　　　　　　7　Spectrum Processing Software

Fig. 7.31 Control of lines in the line profile

You can turn on "Points Visible" in either the Star Profile or Reference Profile. This shows the individual data points. While this tends to clutter the profiles it does allow seeing precisely where the point is and its associate intensity and wavelength. This is particularly good for identifying peak and minimum points for low-resolution work (Fig. 7.32).

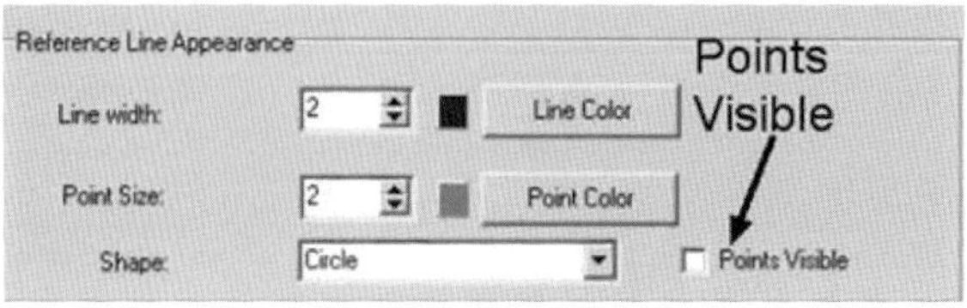

Fig. 7.32 Points visible

Graph

The right most tab at the upper left of the Appearance Window is the "Graph" tab. This is perhaps the most important selection. It allows you to put a title on the line profile, set colors, backgrounds and other options (Fig. 7.33).

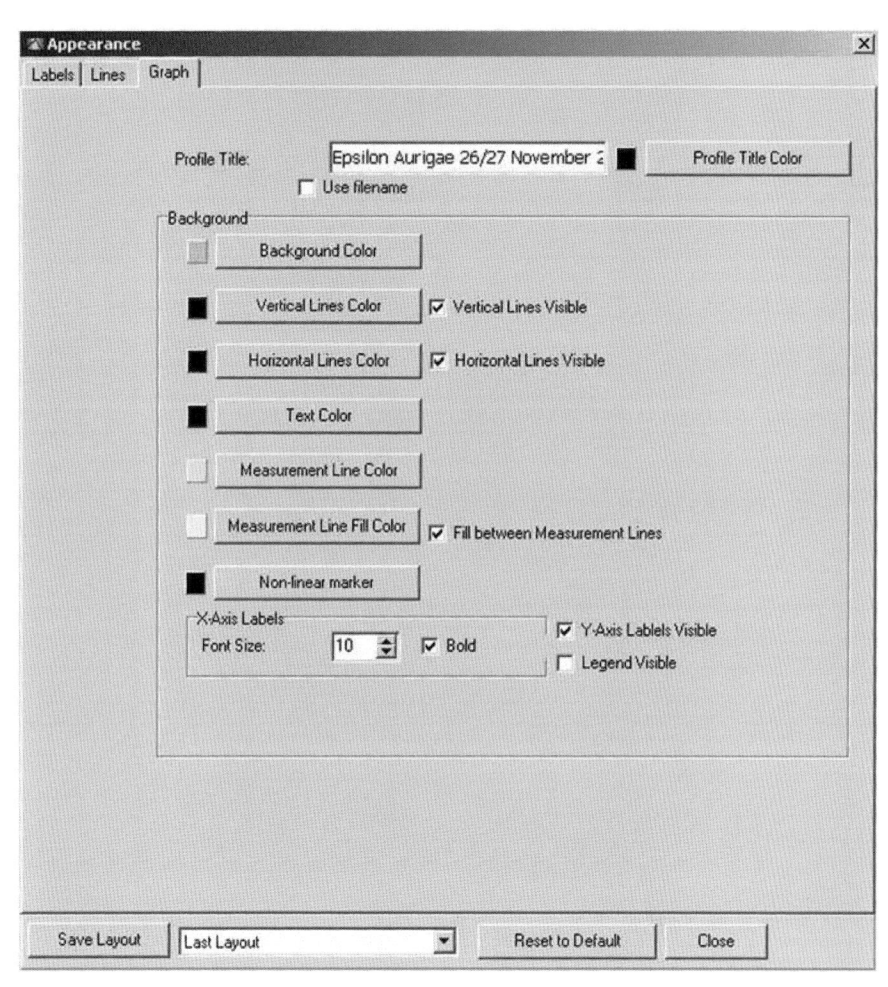

Fig. 7.33 Graph labeling the profile title

To remove the Profile and Reference legend from the upper right of the Profile window, the "Legend Visible" check box can be deselected. This makes the profile wider (Fig. 7.34).

Fig. 7.34 Deselect legend visible

Perhaps the most important part of the Graph selections is the ability to name the line profile.

Reference Menu

Selecting the "Reference" button will display a window with seven options (Fig. 7.35).

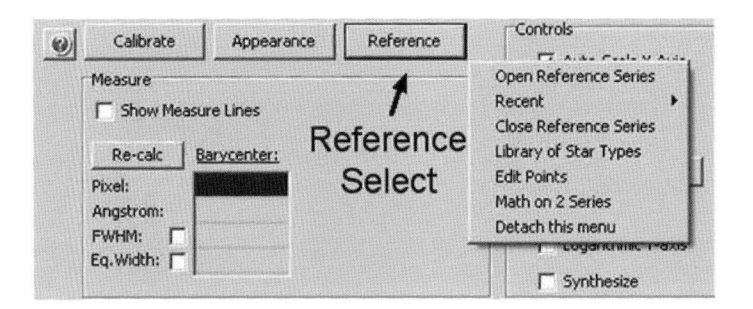

Fig. 7.35 Reference menu

First, because this menu may be used a great deal, there is a "Detach this menu" selection at the bottom that allows the menu to be moved around. The "Edit Points" selection allows the line profile to be trimmed as well as delete unwanted features on the profile, e.g., when creating an instrument spectral response curve (Fig. 7.36).

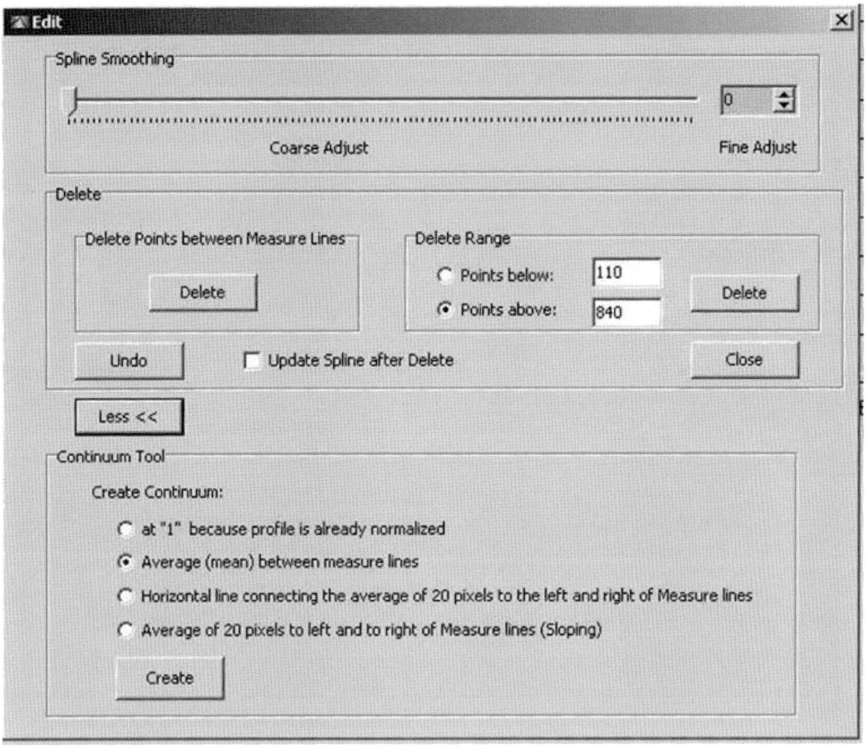

RSpec Spectrum Processing Software

Fig. 7.36 Edit menu

The line profile usually has areas of pixel columns before and/or after the main part. These are of little or no value and can be deleted to expand the profile.

Delete Range

By specifying "Points below" and "Points above" sections of the line profile from the Delete Range menu, the profile can be trimmed and expanded (Fig. 7.37).

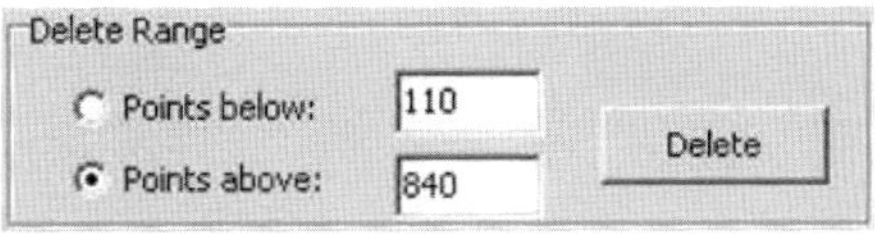

Fig. 7.37 Delete range

200   7 Spectrum Processing Software

Note, when using an un-calibrated profile the points to delete are specified in pixels. A calibrated profile uses points in Ångströms. The following Figure shows points to be deleted in pixels (Figs. 7.38 and 7.39).

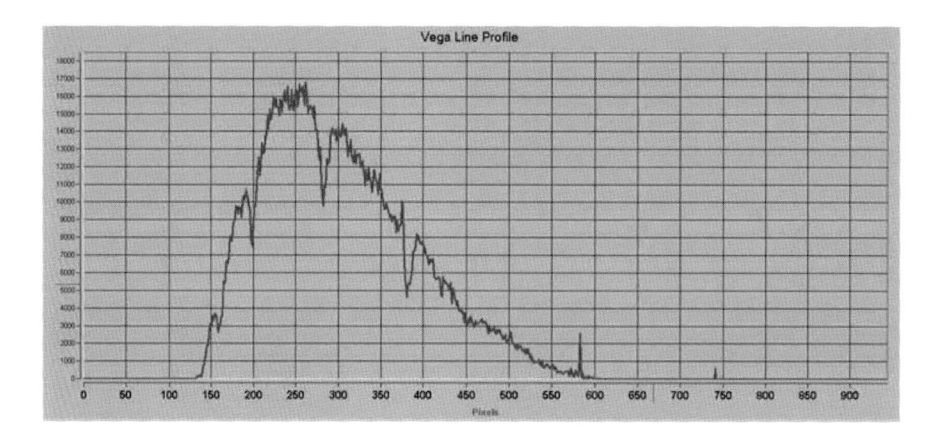

Fig. 7.38 Full image line profile

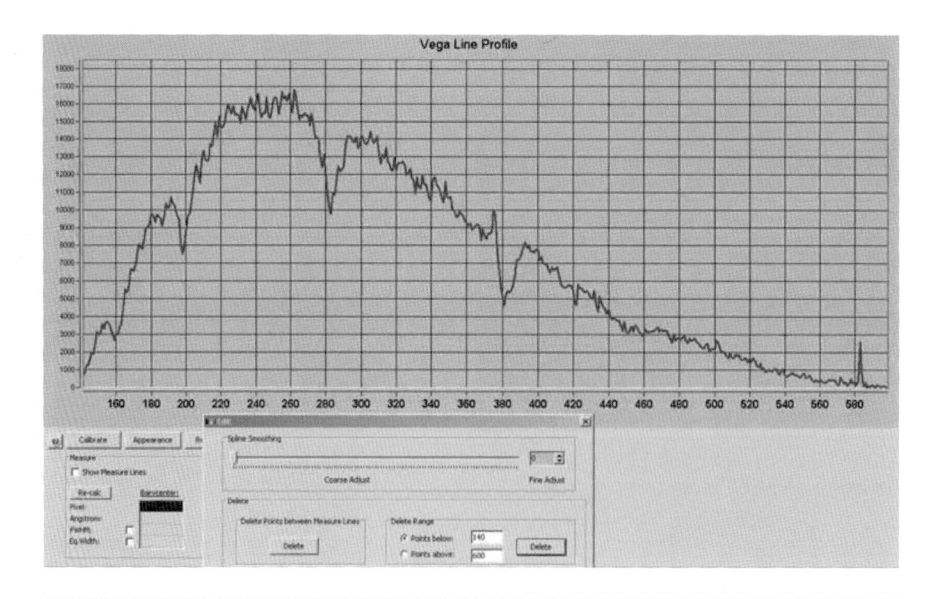

Fig. 7.39 Trimmed/expanded line profile

RSpec Spectrum Processing Software

Delete Points Between Measure Lines

A second way to delete areas before and after the spectrum profile is by using the Measure Lines and "Delete Points between Measure Lines" button.

To use the Measure Lines technique position the lines to bracket the end to be deleted, one line at the end of the graph and the other at the end of the profile where the deletion is desired. Click the "Reference" button and select "Edit Points."

Select "Delete" and the selected end of the line profile pixel columns will be deleted. Next trim the other end of the profile the same way.

Delete Line or Feature

Sometimes, e.g., when creating a response curve, it is desirable to delete certain parts of the line profile without actual shortening the profile. Also there are times when a hot pixel may sneak through or a background star that will produce an unwanted spike. This function can be used to eliminate those unwanted spikes (Fig. 7.40).

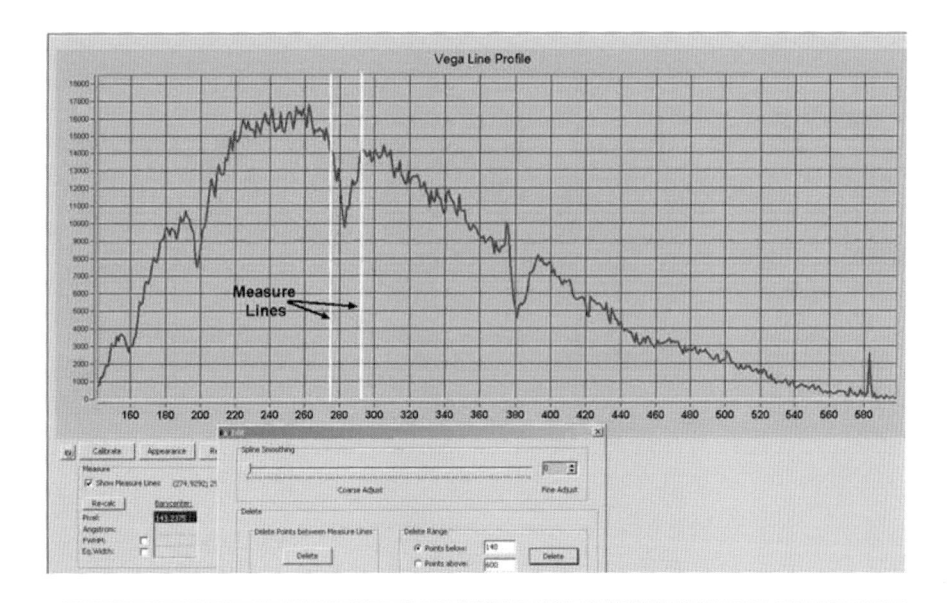

Fig. 7.40 Delete area between white lines

Use the white vertical lines to bracket the area of the line profile to be deleted. When deleted you will notice the area is still there, but where the line feature was there is now a straight line connecting the points where the white Measure Lines intersected the line profile curve. Note, the color of the lines, text and background can be changed from the Appearance menu. The above image is with default values.

The following is an example of an area of a line profile replaced by a straight line. This may seem strange, but will become clearer later when the Response-Calibration is discussed (Fig. 7.41).

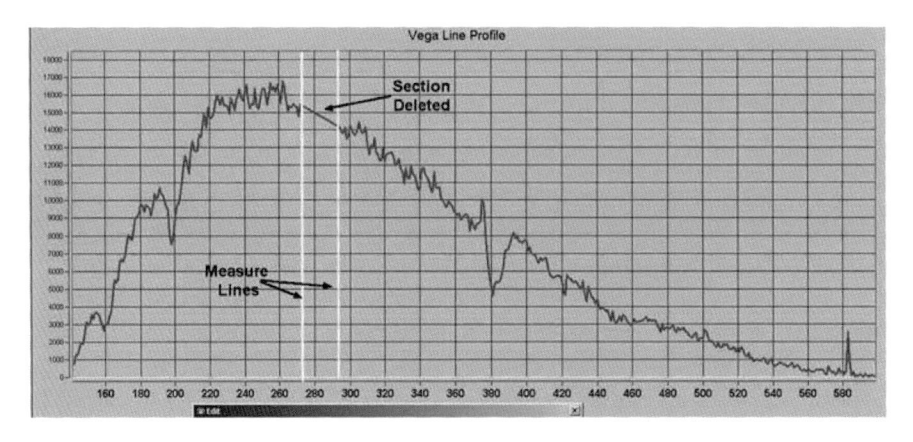

Fig. 7.41 Delete line of line profile

Math

Another very powerful feature is the Math option. The "Math" option is selected from the Reference Menu and the "Math on 2 Series" button (Fig. 7.42).

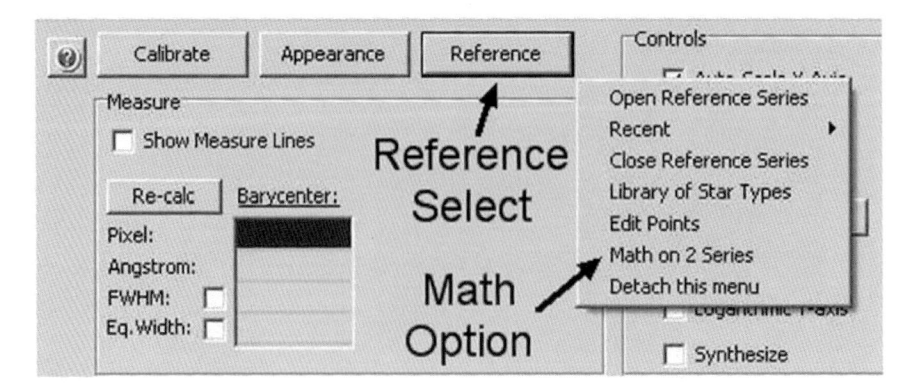

Fig. 7.42 Math option

RSpec Spectrum Processing Software 203

This allows RSpec to perform math functions, e.g., addition, subtraction, division and multiplication of a profile by a reference profile or a number. This is very important for creating the instrument response curve and for normalizing the line profile for equivalent width calculations. This will be discussed in detail later in this chapter (Fig. 7.43).

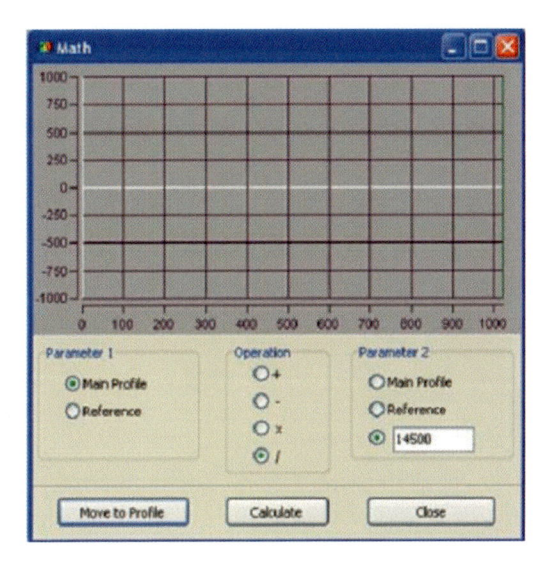

Fig. 7.43 Math menu

Calibrate

This is the Calibrate from the Line Profile Window, not the Math Menu (Fig. 7.44).

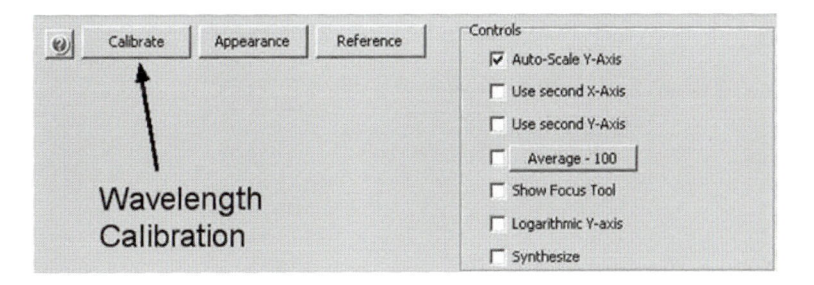

Fig. 7.44 Wavelength-calibration

The Calibrate menu allows you to perform a wavelength calibration on the line profile. There are two basic types of wavelength-calibration, Linear and Non-Linear. For the Linear calibration two points with known wavelengths must be used or a one point with known dispersion (Å/pixel). For low-resolution work with a Star Analyser, if visible, the zero order spectrum can be used as a zero angstrom point. Then just one other point is needed. If an accurate calibration has been previously made and the optical setup has not changed, the determined dispersion along with the zero order (at 0 Å) can be used to provide an easy calibration for the new profile (Fig. 7.45).

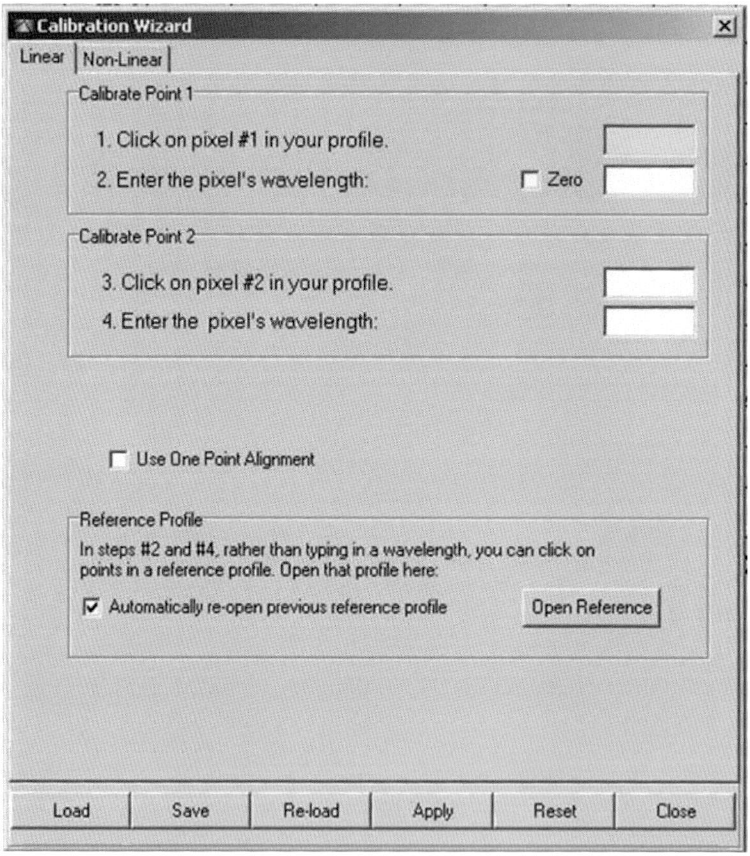

Fig. 7.45 Calibration menu

Measuring Pixel Positions

To be able to wavelength-calibrate a line profile the precise pixel position (X-axis value) of a point or line must be known. There are two ways to do this. The first way is to use the bubble on the graph. By moving the cursor to a point on the profile

RSpec Spectrum Processing Software 205

the pixel position will be displayed (Fig. 7.46). In the following case the position is at X = 218.8378.

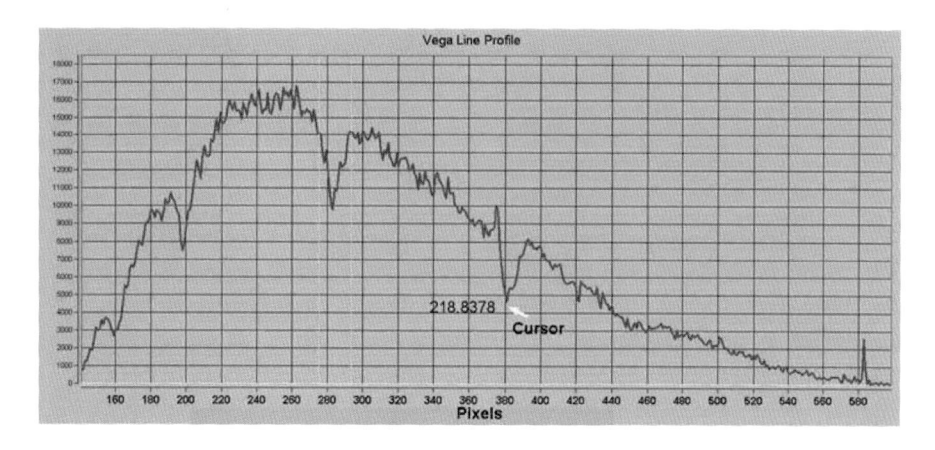

Fig. 7.46 Bubble pixel position

A more accurate position can be determined by using the Measure Lines to bracket the line of interest. The barycenter can then display the precise mid-point pixel position of the line. By bracketing a spectral line with the white Measure Lines the barycenter's precise pixel position can be seen in the area to the lower left. The Measure Line/Barycenter method usually produces a more accurate reading than the bubble method and is required for high-resolution work (Figs. 7.47 and 7.48). In this case the Barycenter position is 222.7421 as opposed to the Bubble position of 218.8378.

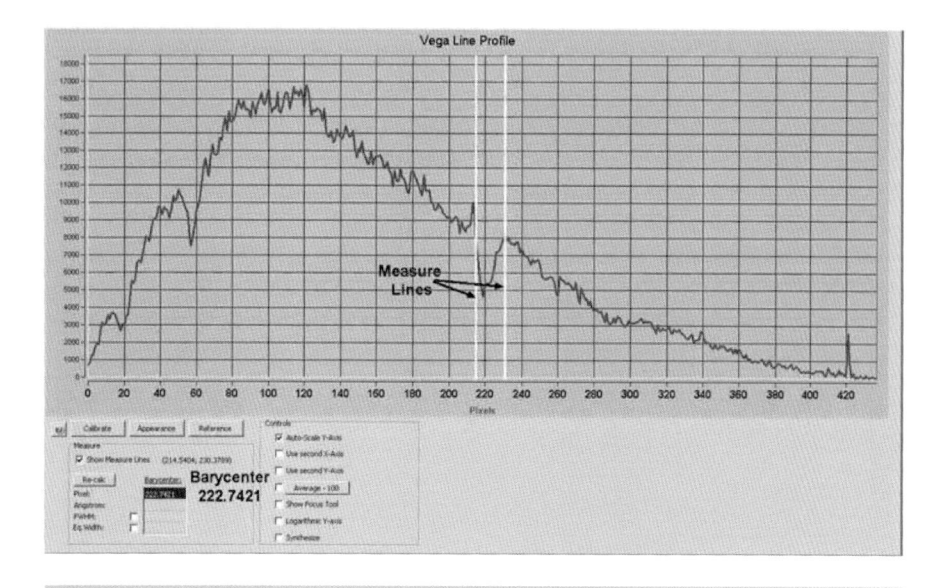

Fig. 7.47 Barycenter pixel position

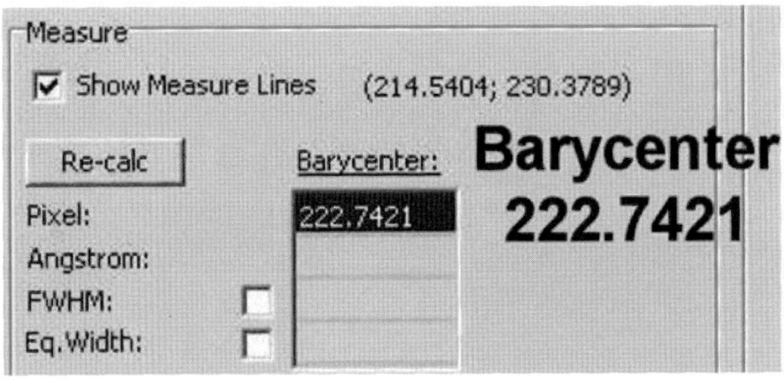

Fig. 7.48 Barycenter detail

Note: Whichever method is used the accuracy is dependent on how accurate the cursor is placed or how accurate the Measure Lines are set.

Wavelength Determination

If the line profile is wavelength-calibrated the Bubble or Measure Lines/Barycenter can be used to show the wavelength at that point. Again the accuracy is dependent on accurate positioning of the cursor or Measure Lines.

Linear Wavelength-Calibration

Selecting the Linear calibration allows one of three methods to be used. These are used mainly for low-resolution wavelength calibration. The first two use two known wavelengths on the line profile. One of these points can be the zero order which is at 0 Å. Just one other point is then needed. Usually the hydrogen alpha line at 6,563 Å, hydrogen beta at 4,861 Å or sodium D line at 5,892 Å can be used as the second point. If the zero order line is not available then any two points on the profile can be used. The following Figures use the zero order spectral line of wavelength 0 Å at pixel position 10 and the hydrogen beta line at wavelength 4,861 Å at pixel position 237 for the calibration. This produces a wavelength-calibrated line profile with a dispersion of 21.4141 Å/pixel. While the dispersion is specified to 4 places in reality just 21 Å/pixel is sufficient for low-resolution work (Fig. 7.49).

RSpec Spectrum Processing Software

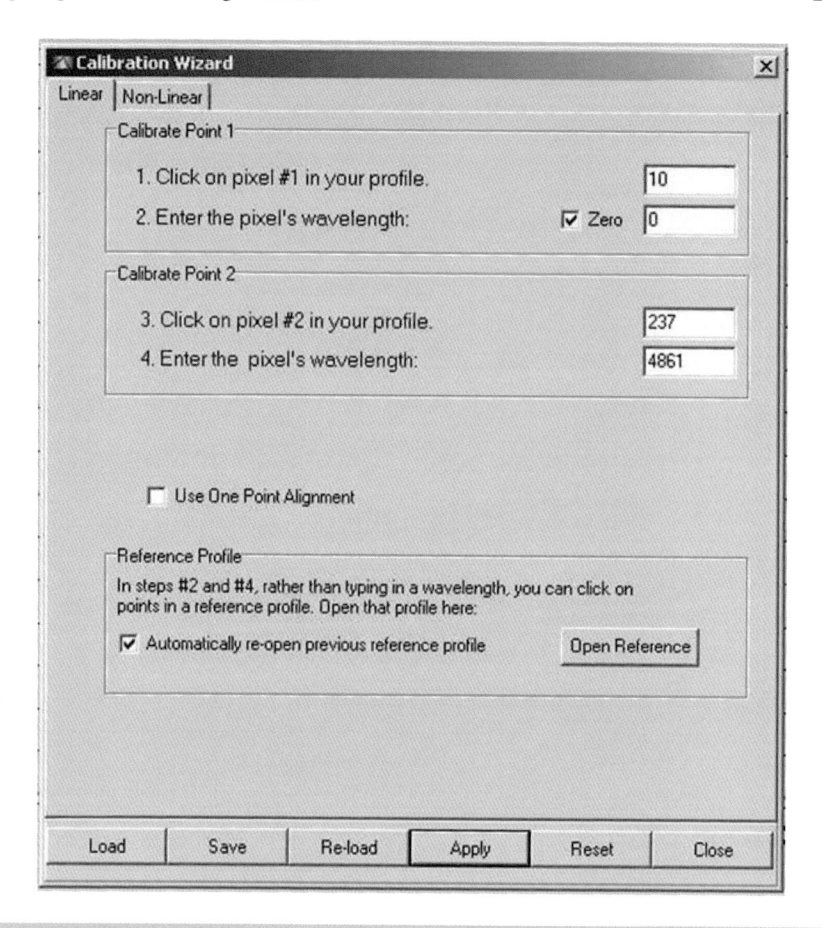

Fig. 7.49 Two point wavelength-calibration

The third method assumes that a wavelength-calibration has been done in the past and the dispersion (Å/pixel) determined. Knowing the dispersion, in this case 21.4141 Å/pixel and one point on the profile, hydrogen beta at 4,861 Å and pixel position 237 a calibration can be done. Note, even the zero order at 0 Å and pixel position 10 could be used with the dispersion to calibrate the profile. It is very important to remember that if the optical system is changed, particularly the distance between the Star Analyser and CCD, the previously determined dispersion in no longer valid (Fig. 7.50).

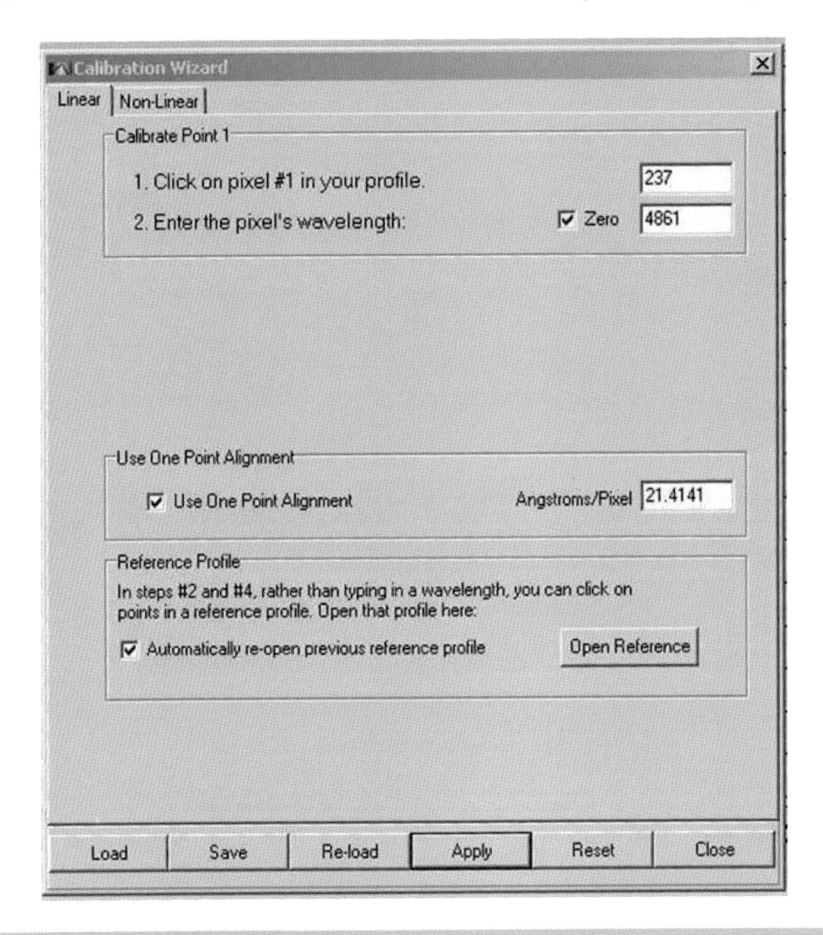

Fig. 7.50 One point wavelength-calibration

Non-Linear Wavelength-Calibration

When doing medium and high-resolution spectroscopy and if more than two calibration points are available, the non-linear calibration will provide a more accurate wavelength-calibration. Selecting "Non-Linear allows you to enter multiple reference points (up to 20) for the wavelength-calibration. High-order calibration can be selected. It is necessary to have at least one calibration point more than the degree of the higher order, i.e., for a 4th order at least five points are needed. In many cases it will be linear rather than non-linear, but if the calibration is indeed non-linear then a higher order calibration curve can be generated. Pixel positions and corresponding wavelengths are entered into the form. When the "Calibrate" button is selected a plot is displayed with the calibration points. A line is drawn through them to show how well they fit. To aid in creating a non-linear curve, higher orders can be selected. As mentioned above at least one more point than the order is needed, i.e., for a 3rd order curve at least 4 points are needed. If five or more points are used a 4th order curve can be generated. Many times the 4th order curve, while correct for the points entered, is not a good curve for other points. Usually the highest order curve that is well behaved is the 3rd order curve. This is an area where the user can easily experiment and see the results on the plotted graph. When using higher orders care must be taken to make sure the data entered is as accurate as possible. If data that are slightly off are entered the higher order will try to produce a fitting curve and will actually make the results worse than with the lower order. When done, clicking on the "Apply" button will wavelength-calibrate the line profile (Fig. 7.51).

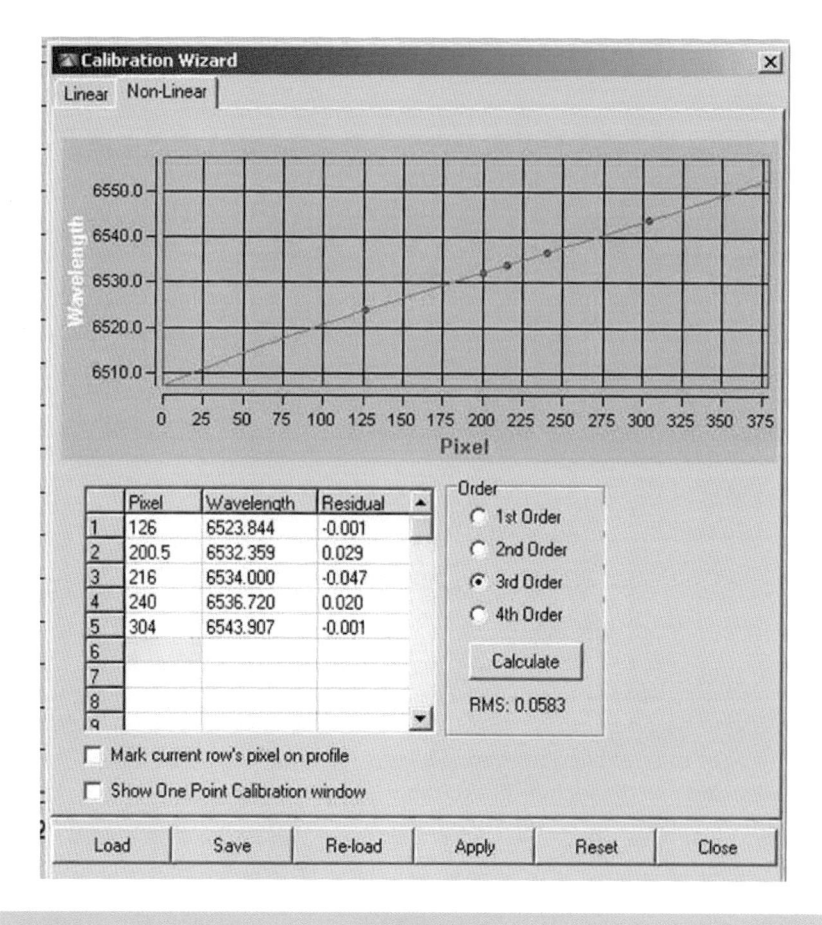

Fig. 7.51 Non-linear window

Elemental Line Check

RSpec has the ability to display elemental lines such as the hydrogen Balmer lines and He lines as well as many other sets of lines on your calibrated line profile. This allows you to see if the elemental lines line up with what should be the corresponding lines on your profile. To show the elemental line menu the "Elements" icon can be selected from the Line Profile Window menu bar. The desired elements or items can then be selected. In the following case the Hydrogen Balmer Lines were selected. These are important as most stars show these to some degree. More than one set of elemental lines can be selected and displayed at one time. Displaying the lines has no effect on the line profile. By selecting different sets of lines, it is possible to identify elements in the spectrum easier (Fig. 7.52).

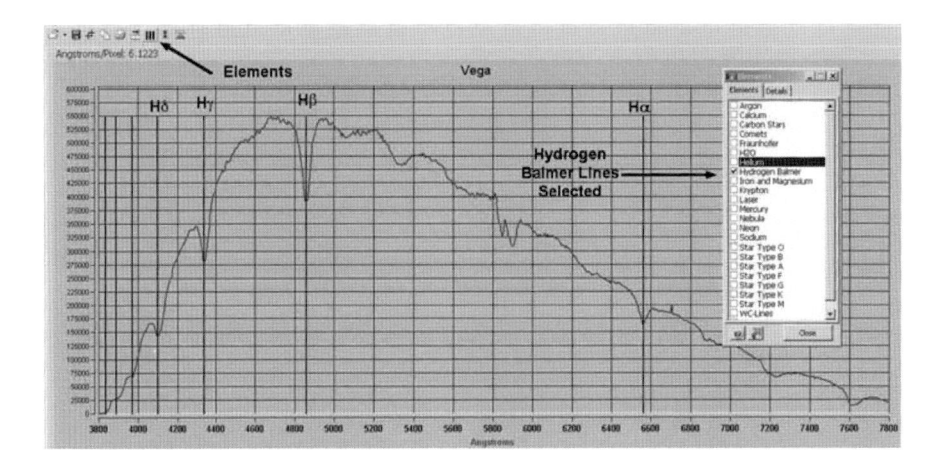

Fig. 7.52 Selecting elemental lines

Data (.dat) File Editing

RSpec uses .dat files in many places. The nice thing about these is you can modify them yourself. Just open them with a text reader, such as Note Pad. In earlier versions of RSpec the Helium lines were not included. This file can be created and the lines added and saved. Some of the versions of RSpec have a Helium file, but lack the He I line. This can be added at the bottom by just typing the wavelength space 1 space # and then saving the file.

For example: 6678 1 #

Along the same lines other Element files can be edited or new custom files created. This is a very handy feature.

The following Figure shows the Helium .dat file opened with Notebook. The last line has just been added for the 6,678 Å line (Fig. 7.53 and Table 7.1).

RSpec Spectrum Processing Software   211

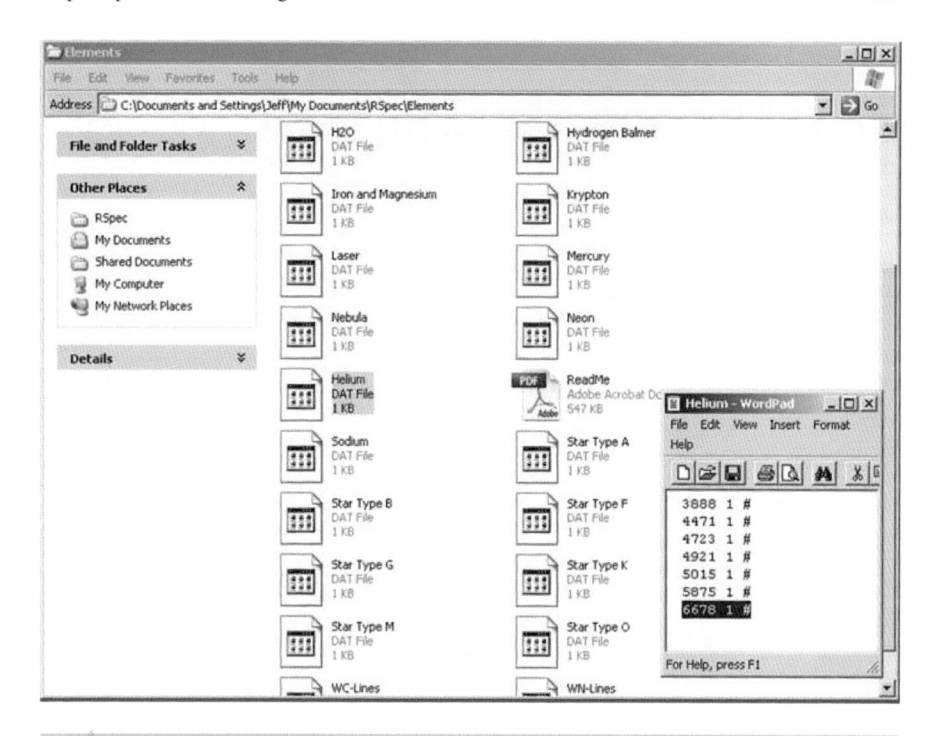

Fig. 7.53 Data file editing

Table 7.1 Select element wavelengths

The hydrogen Balmer line wavelengths are	
Hα	6,563 Å
Hβ	4,861 Å
Hγ	4,340 Å
Hδ	4,102 Å
Hε	3,970 Å
The helium line wavelengths are	
He I	6,678 Å
	5,875 Å
	5,015 Å
	4,921 Å
	4,723 Å
	4,471 Å
	3,888 Å
The average sodium D line wavelength is	
Na Davg	6,563 Å

Instrument Spectrum Response Correction Procedure

A CCD camera and other optics comprising a spectrograph system will have different responses and sensitivities at different wavelengths. While it is sometimes desirable to correct for this system spectral response, many if not most of the time, it not necessary to bother with this correction. All the standard library star line profiles have been corrected. This is the reason uncorrected line profiles do not look like the library standards. The response correction is done mainly for the low-resolution spectra. The reason this is less important with high-resolution work is because the spectrum window is very small compared to the low-resolution image and thus only a very small part of the spectrum is being examined. However, it may be worthwhile to do the correction on high-resolution line profiles for some circumstances.

The following is a description of a procedure for producing a wavelength sensitivity (response) calibration curve that can be used to divide into a spectrum's line profile and correct for the system's wavelength sensitivity. Most any star can be used for this, but a type A has benefits of having spectral lines easily identified to produce a wavelength calibrated line profile. A wavelength-calibrated line profile is needed for the production of the wavelength sensitivity calibration curve. Because several different curves will be produced, rather than try to give them meaningful names alphabetic letters will be assigned the curves to keep things simple.

Note: Computer software does not handle division by zero well. If a line profile's ADU counts go to zero and that profile is divided into another profile, the division by zero may cause problems. If this happens, trim the profiles so none of the profile's pixels go to zero ADU counts.

Response Calibration Procedure

1. Load the star spectrum and preprocess the spectrum image.
2. Create a line profile of the spectrum.
3. Trim the line profile.
4. Wavelength-calibrate the line profile.
5. Note the upper and lower wavelength limits (for this case, 3,800 and 7,600 Å) (Fig. 7.54).

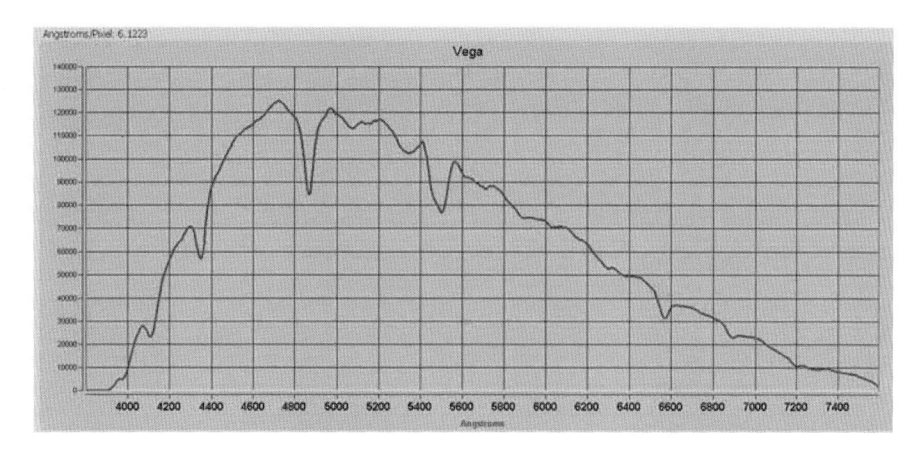

Fig. 7.54 Program star line profile trimmed and calibrated

6. Use the Measure Lines to delete significant lines and features and then create and smooth the curve (Fig. 7.55).

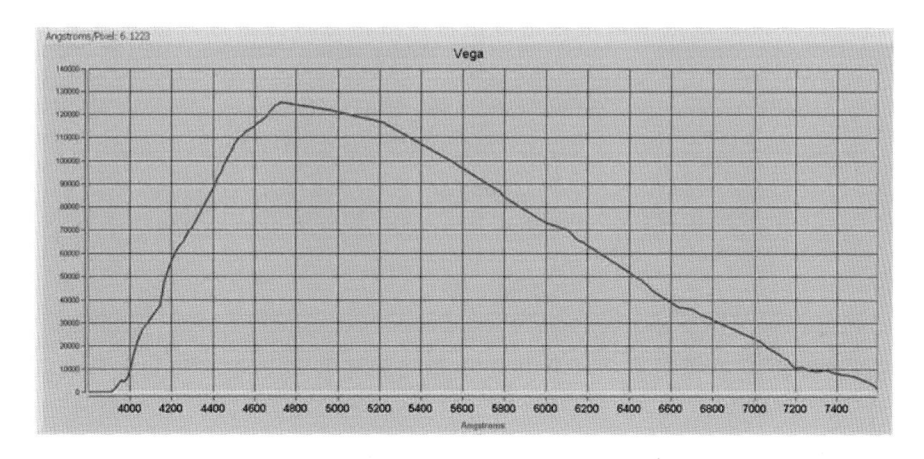

Fig. 7.55 Program star line profile smoothed

7. From the "Edit Points" menu use the "Spline Smoothing" function to produce a second curve. Adjust the "Spline Smoothing" to get the best fit to the smoothed Program Star line profile (Fig. 7.56).

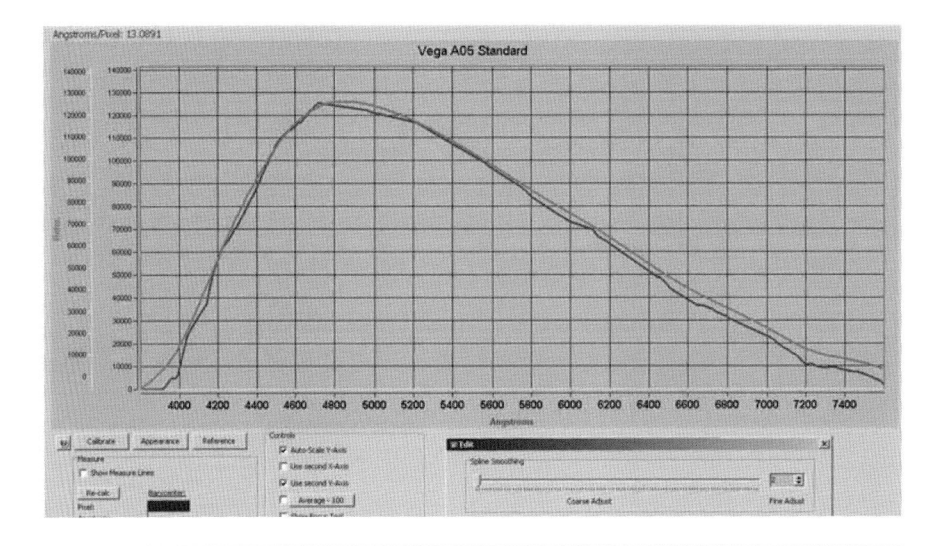

Fig. 7.56 Program star line profile splined

8. Save the Splined Curve as a "Main Profile" called Curve A (Fig. 7.57).

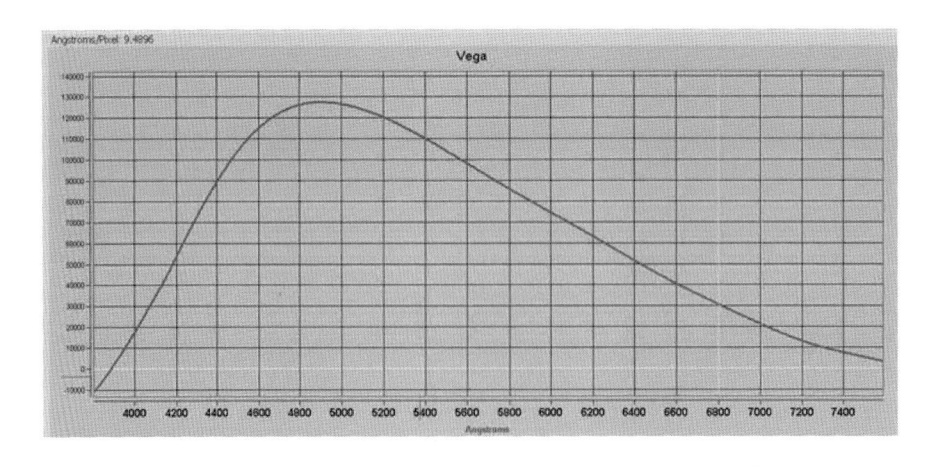

Fig. 7.57 Curve A

9. Load a standard star spectrum line profile of the same type (For this example use type A05).
10. Trim the line profile (Fig. 7.58). To the same upper and lower wavelength limits as curve A (3,800 and 7,600 Å).

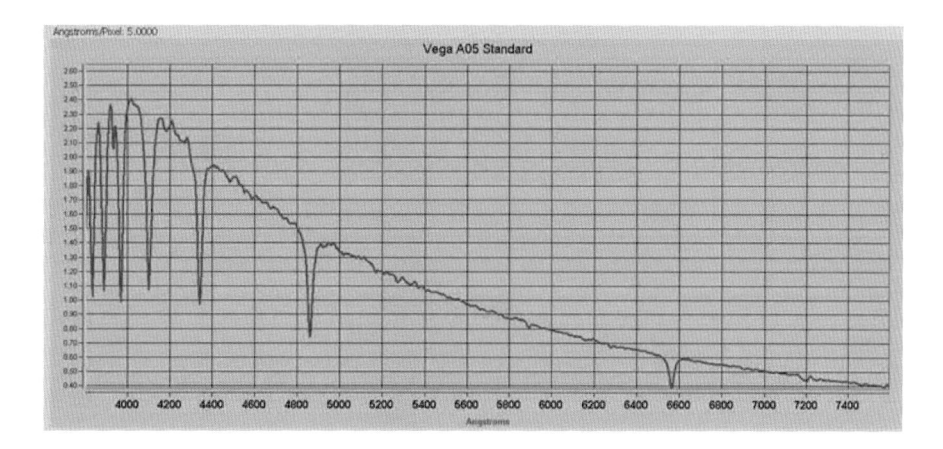

Fig. 7.58 Standard star line profile (A0V.DAT) trimmed

11. Delete significant lines from the profile using the Measure Lines (Fig. 7.59).

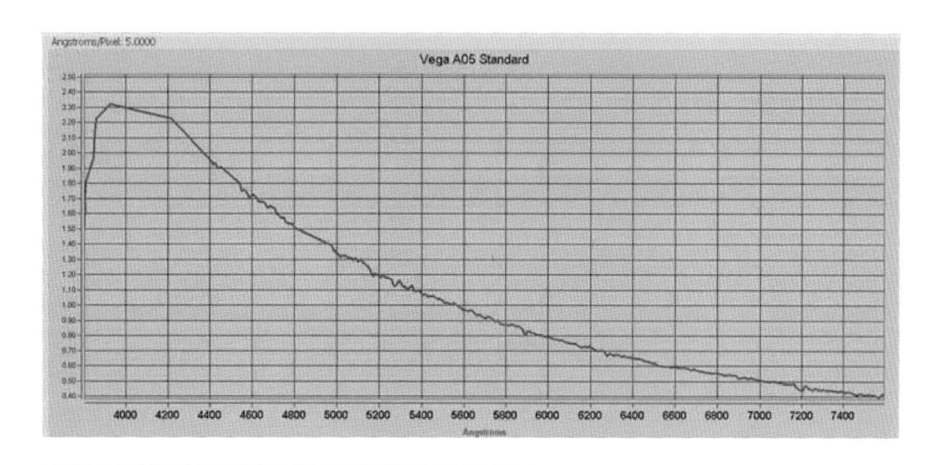

Fig. 7.59 Standard star smoothed

216   7 Spectrum Processing Software

12. Again from the "Edit Points" menu use the "Spline Smoothing" to produce a second curve. Adjust the "Spline Smoothing" to get the best fit to the smoothed Standard Star line profile (Fig. 7.60).

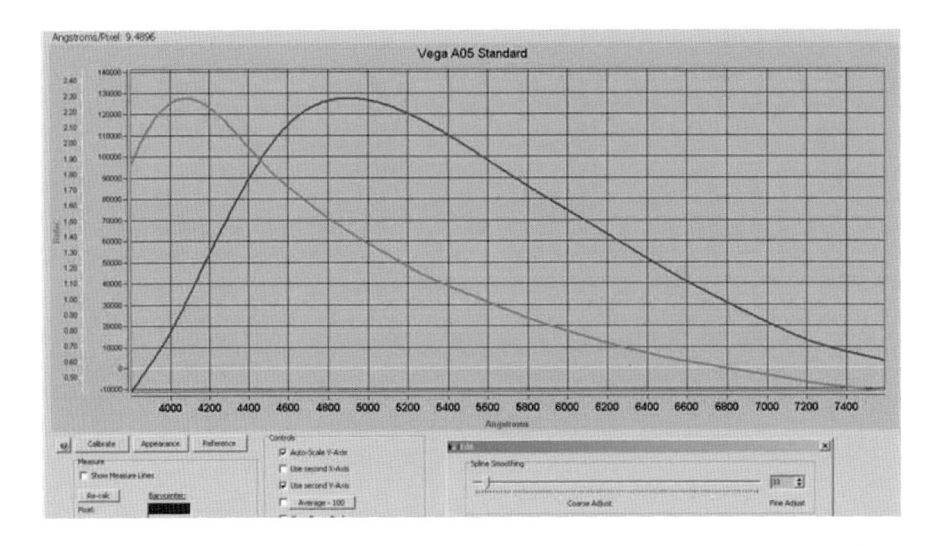

Fig. 7.60 Standard star line profile splined

13. Save the Curve as a "Reference" called Curve B (Fig. 7.61).

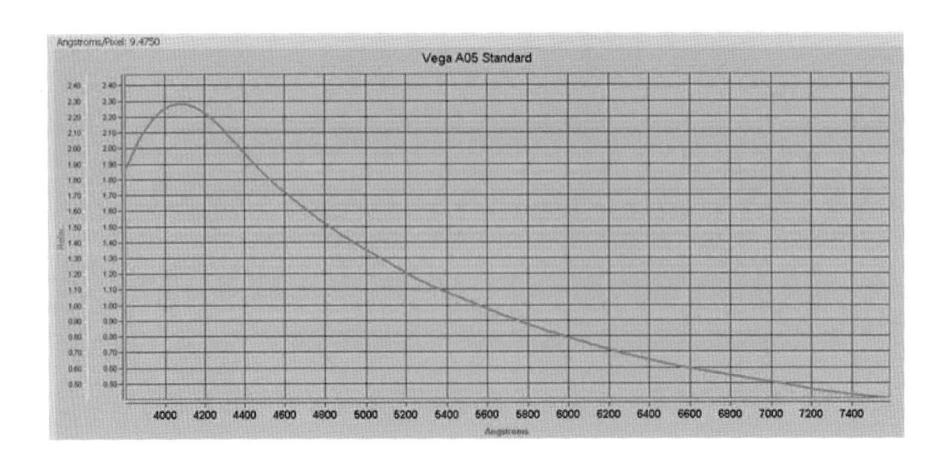

Fig. 7.61 Curve B

RSpec Spectrum Processing Software 217

14. Load the Main Profile Curve A from the Load icon at the upper right of the Profile window and Reference Curve B from the "Reference" button and "Open Reference Series" selection (Fig. 7.62).

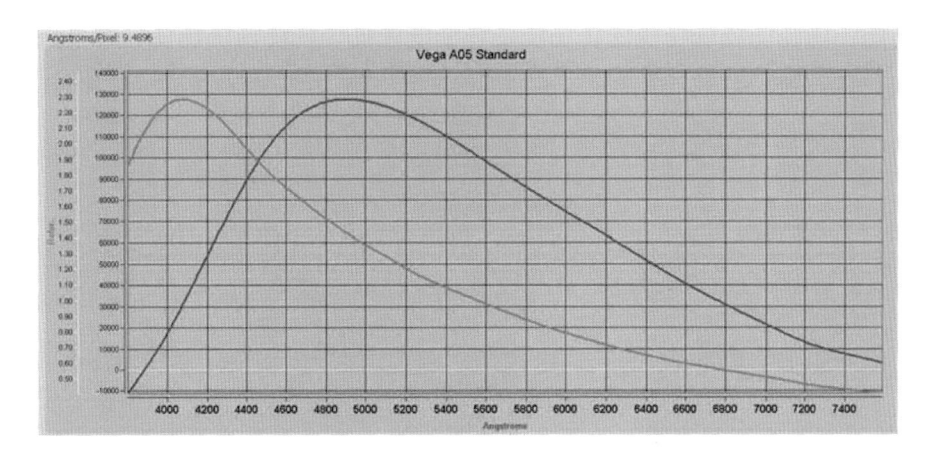

Fig. 7.62 Curve A and curve B

15. From the "Reference" button and "Math on 2 Series" divide Curve B (Reference) into Curve A (Main Profile) creating Curve C (Fig. 7.63).

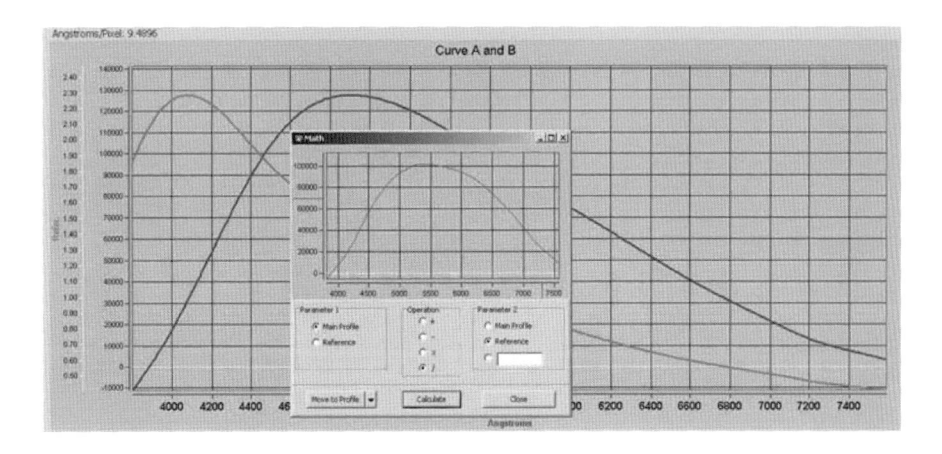

Fig. 7.63 Math series curve A/curve B

16. Move Curve C to the Line Profile window (use the Move to Profile button) (Fig. 7.64).

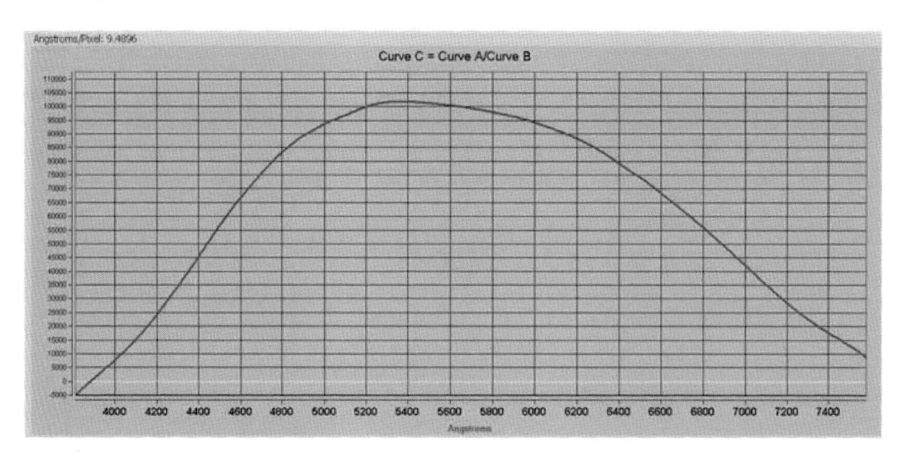

Fig. 7.64 Response calibration curve C

17. This is the Instrument Spectrum Response Curve.
18. Save it as a Reference called Curve C.
19. Load the star spectrum and preprocess it.
20. Trim the line profile.
21. Wavelength-calibrate the line profile.
22. Load the Reference Curve C (Fig. 7.65).

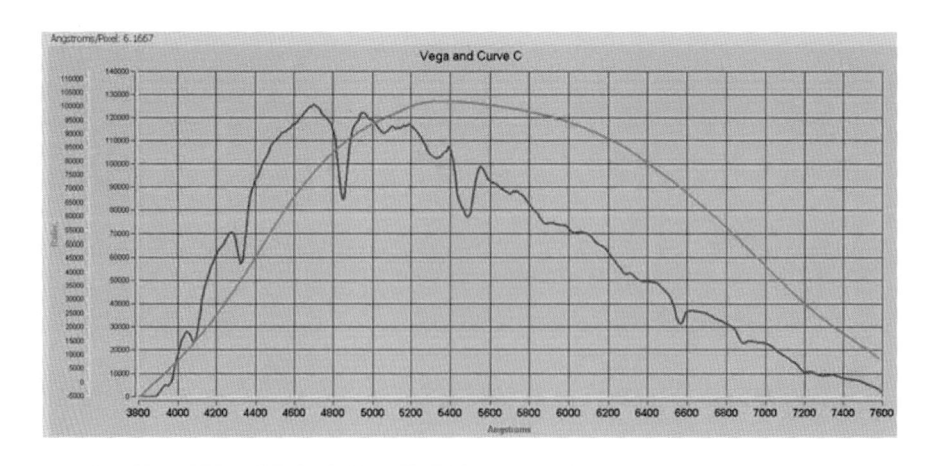

Fig. 7.65 Program line profile and calibration curve C

23. Divide (Math) Curve C (Reference) into the Main Line Profile (Fig. 7.66).

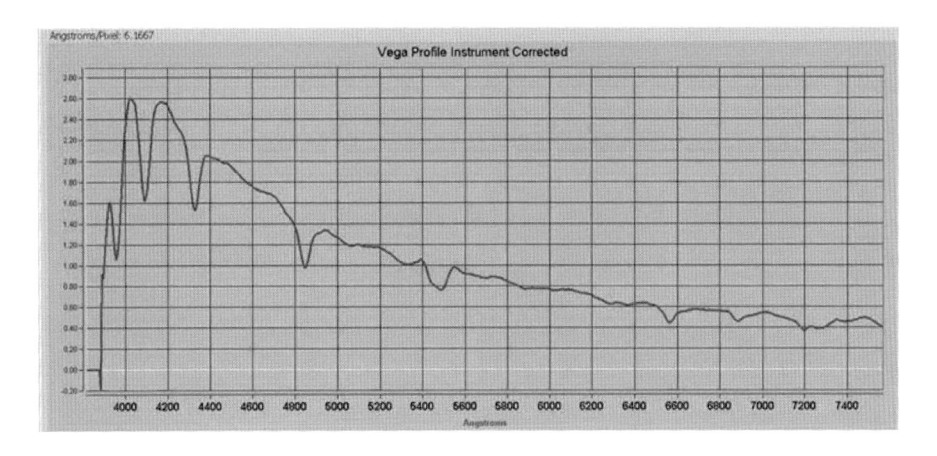

Fig. 7.66 Response calibrated line profile

Synthesize

An interesting feature of RSpec is Synthesize function. By selecting this function a pseudo spectrum of a wavelength calibrated line profile will be created. This can be displayed in black and white or in color. The color and monochrome contrast can be varied with the "Syn Adjust." To use this function, at the bottom of the "Controls" window, select "Synthesize." The icon to the right of the synthesized spectrum is the "Syn Adjust." This controls the colors and intensity of the synthesized spectrum (Fig. 7.67).

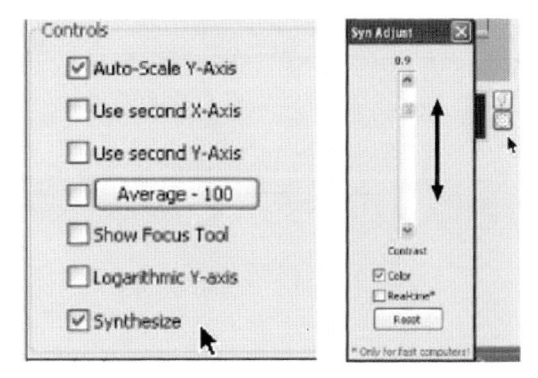

Fig. 7.67 Synthesize option and control

The following Figure shows a low-resolution pseudo spectrum of the star Rigel's line profile (Fig. 7.68).

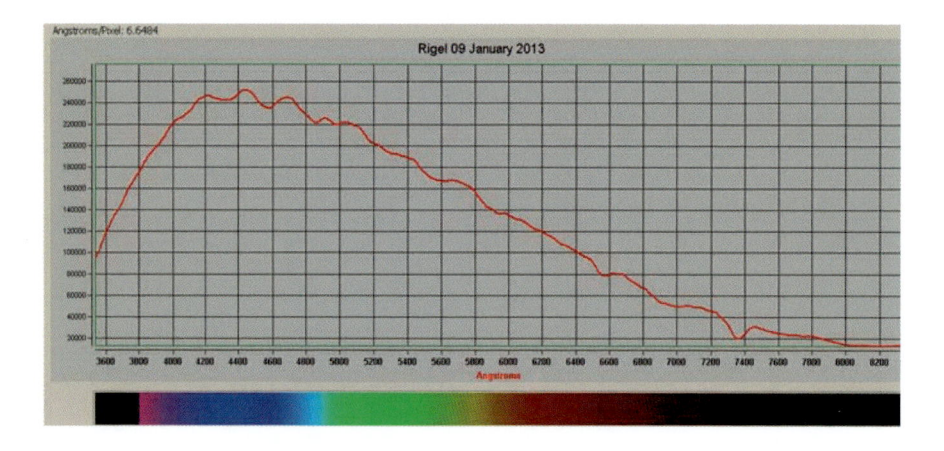

Fig. 7.68 Synthesized low-resolution pseudo spectra

The following Figure shows a high-resolution pseudo spectrum line profile of the hydrogen alpha and sodium D line regions of the star epsilon Aurigae (Fig. 7.69).

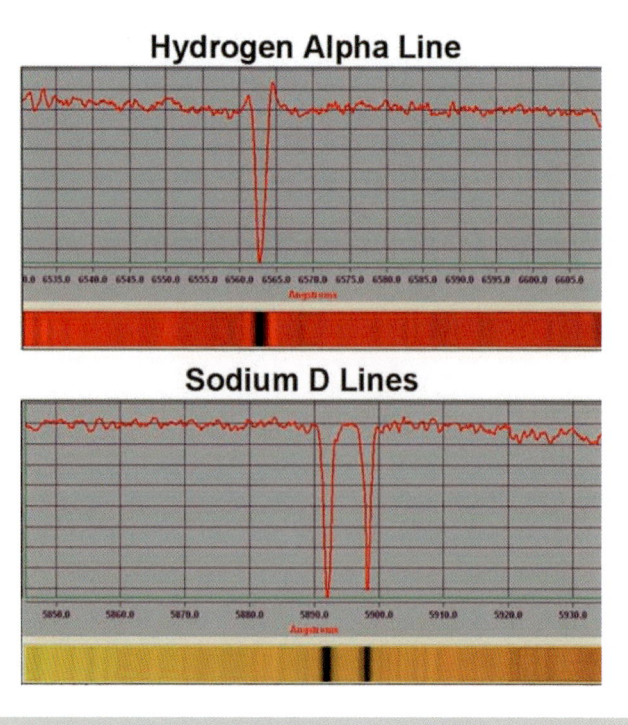

Fig. 7.69 Synthesized high-resolution pseudo spectra

Normalization of the Line Profile Continuum

While RSpec has is no Normalization function, once the average flux (ADU count) value of the continuum in the region of interest is known, that value can then be divided into the line profile using the Math feature to produce a normalized line profile with the spectrum's continuum now set at a value centered around 1.00 (Fig. 7.70).

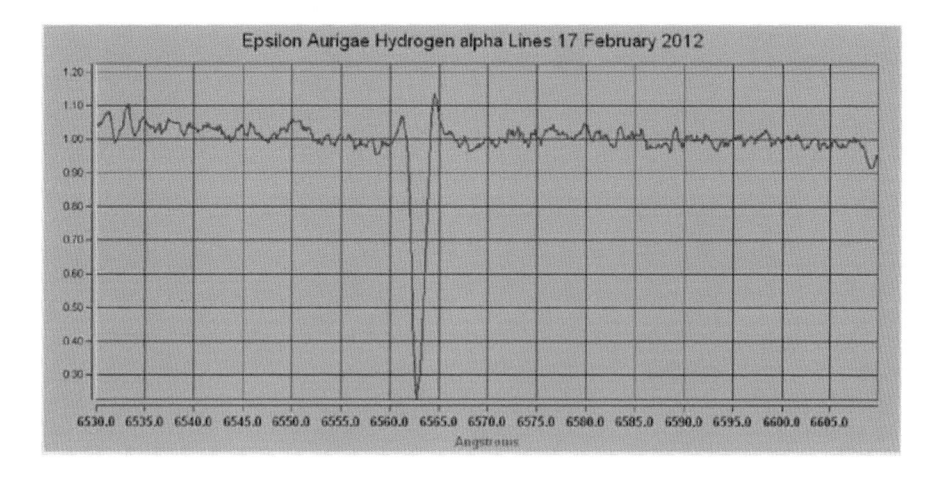

Fig. 7.70 Hydrogen alpha normalized line profile

Equivalent Width Determination

The equivalent width (EW) of spectral lines can be determined once a line profile that has been wavelength-calibrated and normalized. EW is the power or strength of the line as determined by its area. The Measure Lines are set to bracket the line where the line intersects the continuum on the right and left side of the line. Selecting the "Eq. Width" box will then display the EW value in Ångströms (Fig. 7.71).

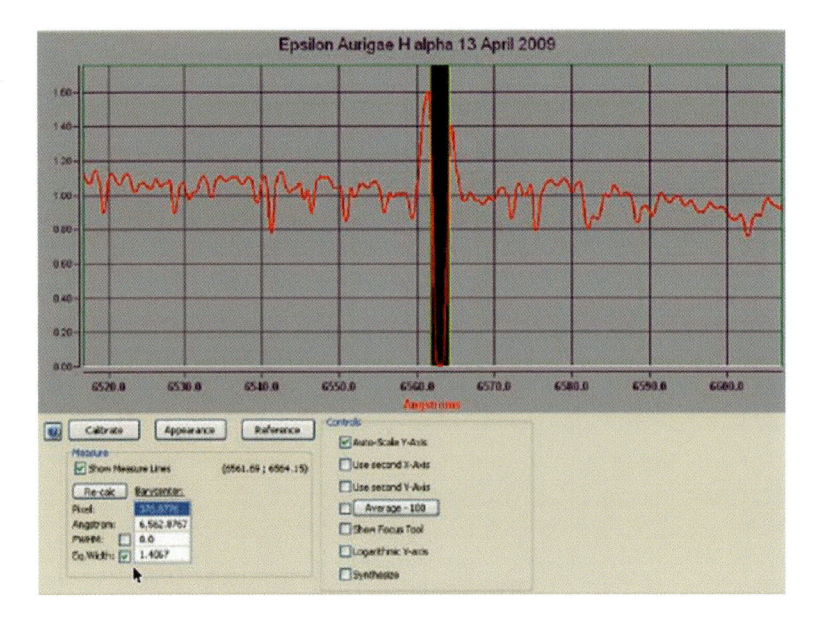

Fig. 7.71 Equivalent width measurement

Heliocentric Calibration

For accurate wavelength determination, a heliocentric wavelength correction should be made. RSpec does not have this feature, but VSpec does. VSpec requires an image to be loaded and a line profile created. Then from the Spectroscopy menu at the top of the screen "Heliocentric Correction" selected. After entering the required data select "Compute" and the program will show a window with the heliocentric correction and several other data items. More information on this will be presented in the VSpec section.

VSpec French Freeware

Introduction

VSpec is a freeware program originally created in French and converted to English. There are detailed tutorials available on the web site. Be aware, this program, while free, requires a steep learning curve and is not the best program for beginners. To use the program effectively it is best to have a firm idea of what you want to do.

VSpec French Freeware

It may take some experimentation and frustration to achieve what is desired. Many of the screens and menus are not seen until something else is done first. This tends to make exploring difficult. Selecting the wrong icon or menu item can cause the program to crash. It is also very easy to get lost. If something is done that was not what was wanted, the results sometimes look reasonable, but in reality are not what was wanted. It is suggested that a simple A star's spectrum, like Vega, be used to experiment with and get familiar with VSpec before trying to tackle more complex spectra. The newer version seems more stable than earlier versions, but it is still fairly easy to crash the program. VSpec has some very powerful features if one takes the time to master it.

The latest version as of February 2013 is 4.1.3. VSpec can be downloaded at: http:/www.astrosurf.com/vdesnoux/index.html

Setting Up VSpec

The opening window for VSpec is a mostly blank yellow page. Until an image file has been loaded there will be practically no other icons or menus available. To the upper left of the window an icon can be selected to open the image file (.jpg, .bmp, .fit) or a profile (.dat) file can be opened with the third icon from the left (Fig. 7.72).

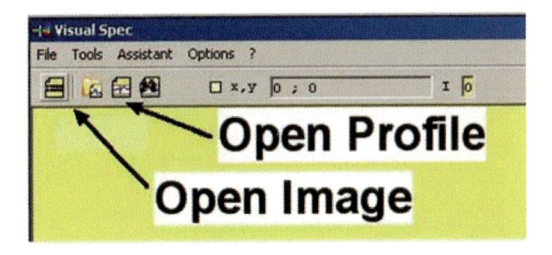

Fig. 7.72 VSpec opening window select fits file

224   7   Spectrum Processing Software

VSpec can open image type files qmps(*.pic), .fit, .fits, .fts and .bmp. If you are using .fits files, be sure to change the "File Type" in the "Open" window to .fits. Otherwise, the .fits files will not be seen. Once a spectrum is displayed the cursor can be moved onto the spectrum to inspect the intensity of the pixels at various points. Both the cursor X/Y position is noted and the ADU count (intensity) at that cursor is shown at the top of the screen (Fig. 7.73).

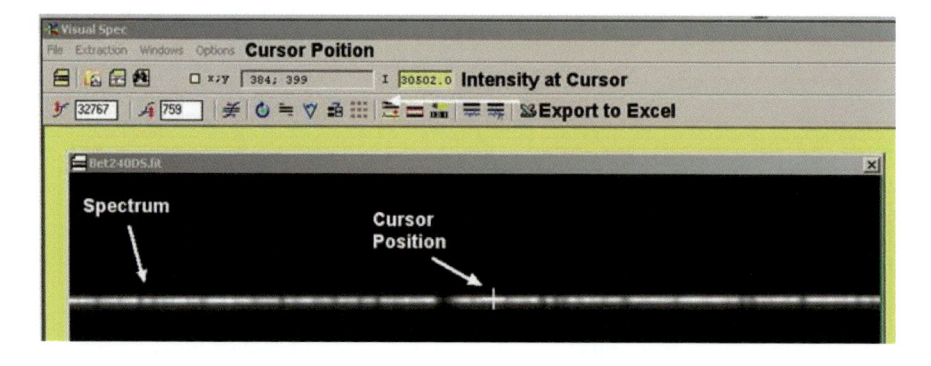

Fig. 7.73 VSpec with spectrum image

Once a spectrum image file has been loaded the menus at the top of the screen change and many new icons shown. The icons may be hard to figure out (Fig. 7.74). Figure 7.75 explains what they are.

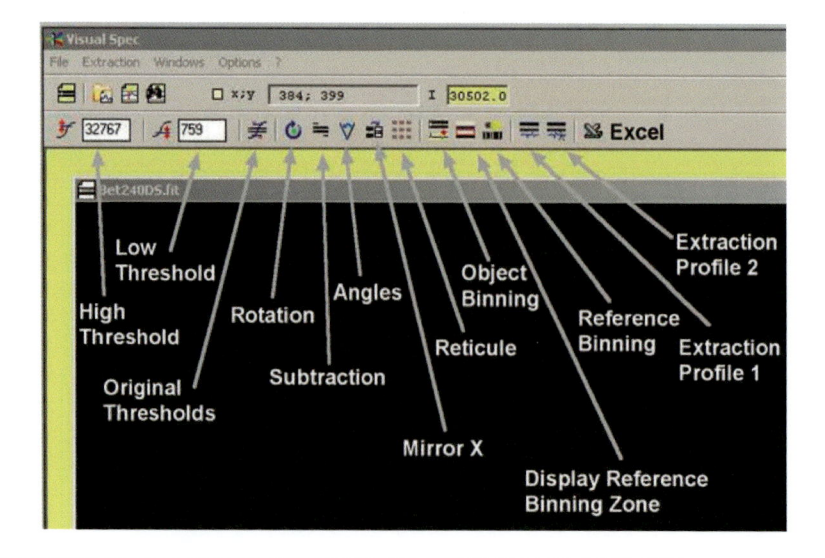

Fig. 7.74 VSpec spectrum icons

VSpec French Freeware   225

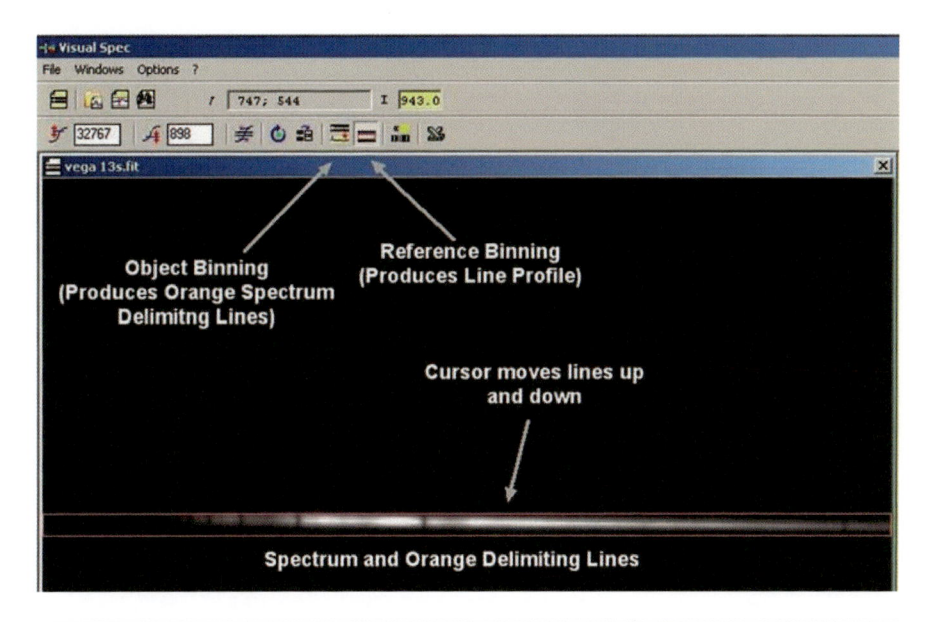

Fig. 7.75 Spectrum delimiting

Selecting the Spectrum

After the spectrum image has been loaded, select the "Display Reference Binning Zone" icon and two orange lines will be shown in the spectrum window. These lines must be moved to bracket the spectrum. Putting the cursor between the lines moves them together. Putting the cursor above or below a line will move just that line.

Line Profile Creation

Once the spectrum has been delimited with the orange lines, select the "Reference Binning" icon and a line profile will be seen (Fig. 7.76).

226   7 Spectrum Processing Software

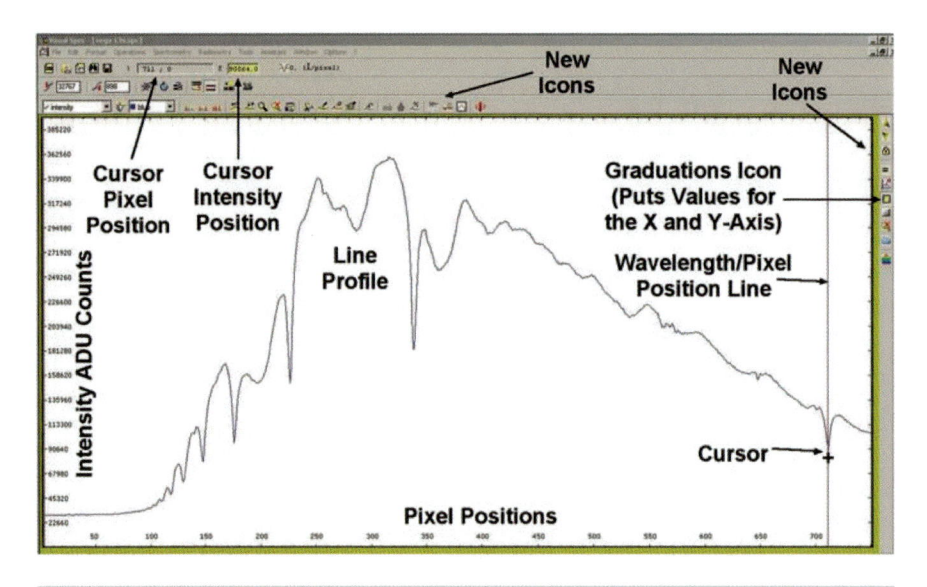

Fig. 7.76 Spectrum line profile

When the line profile is created there will be no X-axis and Y-axis information, but a new set of icons will appear along the right edge of the screen. To put the axis information on the plot, select the "Graduations" icon to the right (Fig. 7.77).

Fig. 7.77 Graduations icon

Wavelength Calibrating the Spectrum

Wavelength calibrating a line profile in VSpec is very similar to the technique used in RSpec and other spectrum processing software programs. The positions of at least two pixels must be known with corresponding wavelengths. The more pixel positions used, the more accurate the calibration will be. For low-resolution work known lines in a spectrum can usually be used unless they have been Doppler shifted a great deal (e.g., with Quasars). For mid and high-resolution spectroscopy a fixed elemental set of spectral lines is used. Sometimes atmospheric telluric lines can be used.

The multiple-line calibration will be used for this discussion. From the "Spectroscopy" tab select "Calibration multiple lines…." The Non linear calibration window will be displayed (Fig. 7.78).

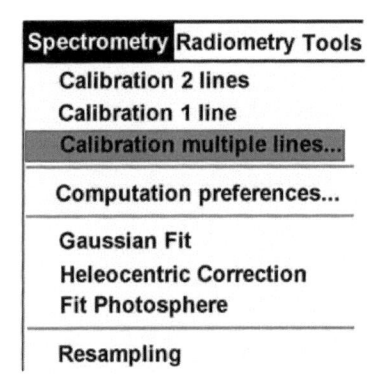

Fig. 7.78 Selecting multiple lines calibration

To use this the element of the calibration lines must be known. In this case the element hydrogen will be used. From the top menu bar the "Tools" tab is selected and the "Elements" window is displayed (Fig. 7.79).

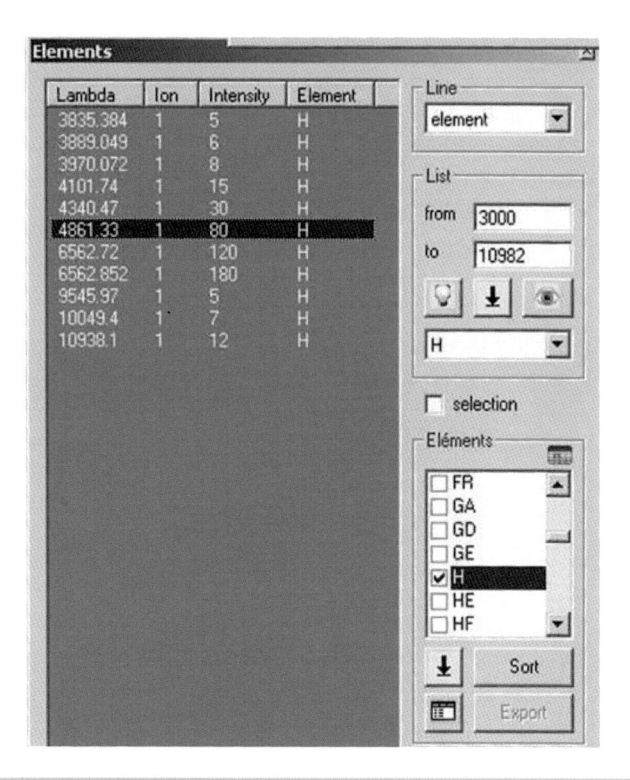

Fig. 7.79 Selecting elements

At the lower right there is a list of elements. Scroll to H (hydrogen) and check the box. Next click on sort. At the left will be a list of the hydrogen Balmer lines and their respective wavelengths. Click on the "4101.74" line in the "Elements" window. Now place the cursor at the left of that line on the line profile and while holding the button down drag the cursor to the right so that the dotted lines are on both sides of the line and the line is centered. A box will appear next to the cursor with the wavelength (4101.74). Hit Enter and the wavelength and pixel position will be entered into the Non linear calibration table (Fig. 7.80).

VSpec French Freeware   229

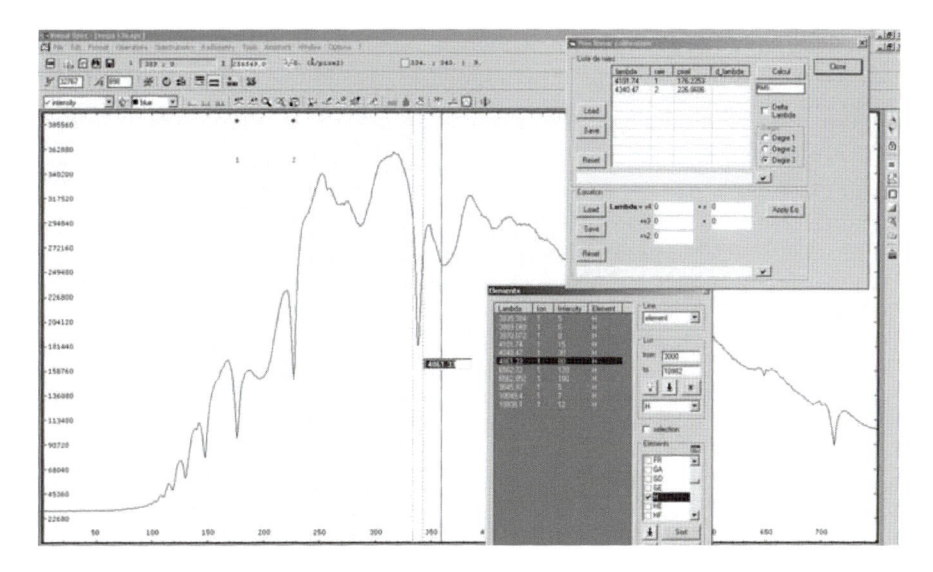

Fig. 7.80 Wavelength selection

Repeat for the rest of the lines. When done, click on the "Calcul" button at the upper right in the "Non linear calibration" window. The profile is now wavelength calibrated and a Dispersion window is displayed to show how well the fit is (Fig. 7.81).

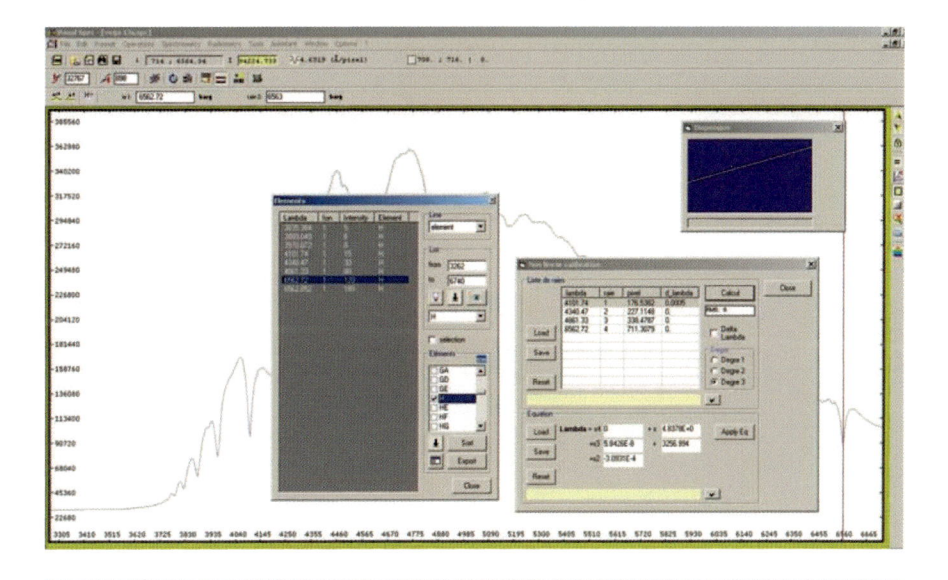

Fig. 7.81 Wavelength calibrated profile

Heliocentric Correction

To make a heliocentric correction for line at a specific wavelength first, from the "Spectroscopy" tab (Fig. 7.82), select "Heliocentric Correction."

Spectrometry	Radiometry Tools
Calibration 2 lines	
Calibration 1 line	
Calibration multiple lines...	
Computation preferences...	
Gaussian Fit	
Heleocentric Correction	
Fit Photosphere	
Resampling	

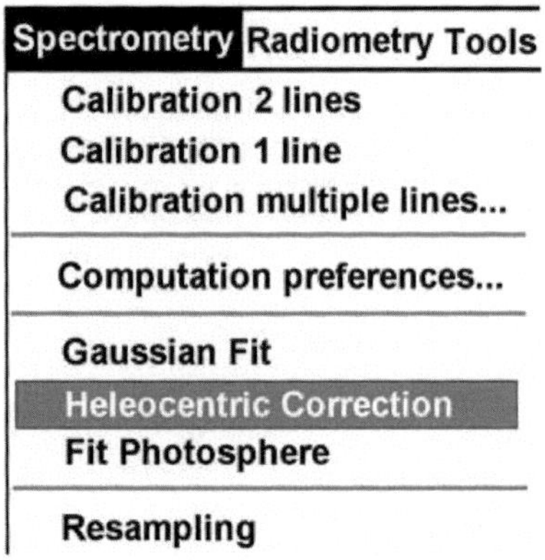

Fig. 7.82 Heliocentric correction selection

A "Heliocentric" window will appear. Enter the observatory's latitude and longitude in decimal form, right ascension (Alpha) and declination (Delta) of the star, day/time in decimal form, month and year and wavelength region of interest (Fig. 7.83).

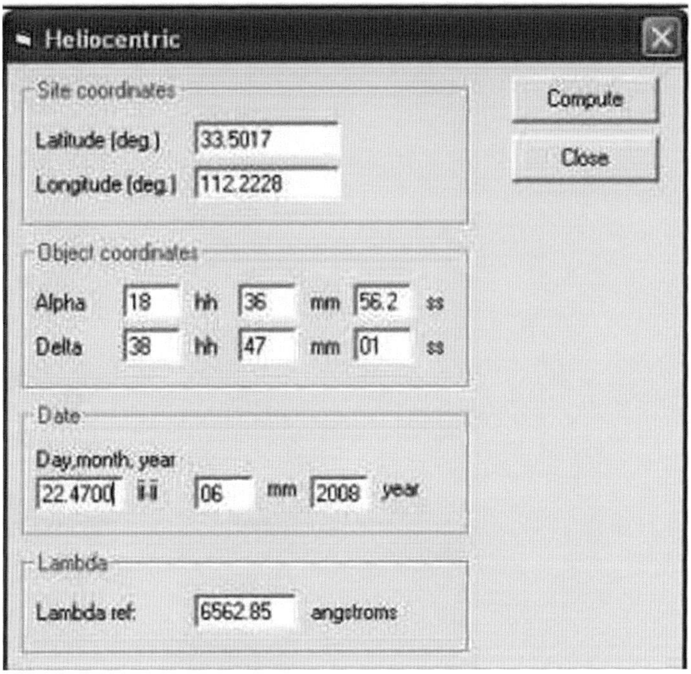

Fig. 7.83 Heliocentric correction menu

The "Compute" button at the upper right is selected and an "Infos …" window displayed. In addition to the heliocentric wavelength correction, the observation's Julian Date along with the radial velocity (Speed), JD Correction, Azimuth, Height (elevation) and Air Mass of the star at the time of the observation will be displayed (Fig. 7.84).

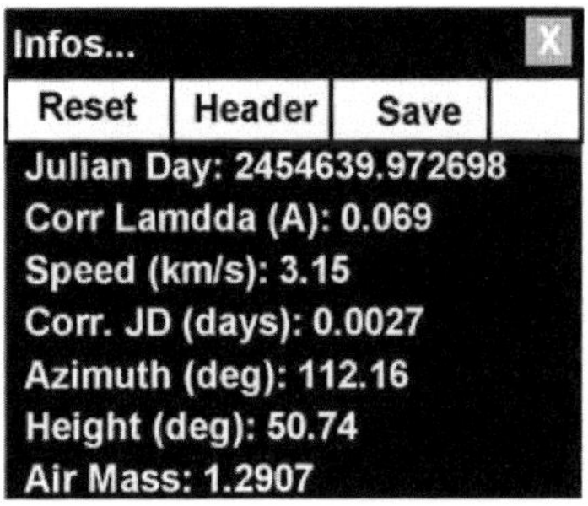

Fig. 7.84 Heliocentric correction results

Chapter 8

Astronomical Spectroscopy Projects

Introduction

Some people feel that it necessary to observe and do spectroscopy on faint stars and stars that have not been observed in detail by others. While this is certainly a worthy goal, it does not work well for small observatories. Because spectroscopy spreads the light out, taking spectroscopic images of faint objects requires very large telescopes to gather enough photons for a spectrum image. Faint astronomical objects are truly in the realm of the large observatories. Small observatory observers (small observatories are observatories typically in the backyard with apertures of 16″ or less) can still make significant contributions by observing bright objects.

Bright star observing has many advantages for the smaller observatories. First, because they are bright, larger observatories do not usually observe them even if they need to be observed. Writing a proposal for time on a large telescope for a bright star is likely to be rejected. Many bright stars desperately need continued spectroscopic observations. A big point for small observatories is that they can observe near continuously.

The following are just a few projects that are well suited for small observatory spectroscopy. The observer is encouraged to research these and similar astronomical objects and set up an observing program. Most of the stars have hydrogen alpha and sodium D line regions that are easy and very interesting.

While spectroscopy can be challenging and fun, to make it even more exciting one can get involved with a Campaign. This has multiple advantages. There is still much to be learned and seeing what others are doing and ways to improve your observing is a great way to progress with spectroscopy. Perhaps the most exciting time is when you can submit data or a paper or make a presentation at a scientific meeting.

Astronomical Organizations

The American Astronomical Society (AAS), the Astronomical Society of the Pacific (ASP), the American Association of Variable Star Observers (AAVSO) as well as the Society for Astronomical Sciences (SAS) have used many presentations and papers from amateur astronomers. For many years these presentations and papers have been dominated by photometric research. Over the past decade, that has changed. Each year there are more papers and presentations by amateur astronomers doing spectrographic research.

An easy way to get involved is with a Poster Paper. If you are a bit shy about getting up in front of a group of people to make a presentation, then Poster Papers may be for you. Both the AAVSO and SAS cater to the amateur astronomer and provide a friendly and fairly inexpensive atmosphere. They have yearly, and sometimes more often, meetings where poster papers and presentations can be made. The AAS and ASP are on a higher professional level and much more expensive.

Stellar Spectroscopy Projects

Introduction

Obviously the easiest star to study is our star the Sun. Because it is so bright, telescopes are not needed and most times a means of reducing the light is needed. While study of the Sun using spectroscopy is interesting and educational, there are several dedicated professional observatories around the world providing continual monitoring of the Sun. Most of the really interesting discoveries have probably already been made and amateur contributions are unlike to produce data of value. Still, it can be inspiring and a learning experience for those interested.

The areas of astronomical spectroscopic most fertile for amateur contributions are stellar and super novae. As note by Dr. John Martin (University of Illinois) "At higher resolution, spectroscopy of any variable star is useful. Relatively little is known about many bright stars because they haven't been as sexy or interesting for most professionals to give attention to. The hydrogen Balmer lines are a good place to start because rotation, pulsation, and star spots affect them. But careful examination would probably identify other lines in variable stars that are sensitive to changes in the star's atmosphere and wind."

Quasars can be studied too, but like the solar spectroscopy, the large professional observatories are best able to push the envelope. But, also like solar observing doing spectroscopy of some of the brighter quasars can be both inspiring and a learning experience.

Stellar Spectroscopy Projects

The following are several projects that have seen amateur spectroscopic contributions and require continued observations. Most of these are bright and easy to observe. Some of the stars are at the end of their life and due to go supernova at any time. Catching spectral changes as a star goes supernova would a great accomplishment. The constellation Orion presents several stars of interest. These are all easy for both high and low-resolution spectroscopy.

Epsilon Aurigae

Epsilon Aurigae Parameters

Spectrum Type: FO Iap, V = 3.0
RA 05 h 01 m 58.1 s, Dec +43d 49′ 24″ (2000)

Introduction

Epsilon Aurigae is a mysterious and bright (V = 3) long-period (27.1 years) eclipsing binary star system. It is easy to find and observe. Epsilon Aurigae is located Northeast of Capella at the vertex of a triangle of stars known as the "Kids." Under dark clear skies it is easily seen with the naked eye (Fig. 8.1). This allows high-resolution spectroscopy with even very modest telescopes and fairly short exposures (in minutes, not hours).

Fig. 8.1 Constellation Aurigae

Epsilon Aurigae Eclipse

The last eclipse of epsilon Aurigae ended in 2010. The next eclipse is in 2036. This does not mean the star should not be studied now and the years between eclipses. There is still much to be learned and the system is constantly changing.

Epsilon Aurigae Star System

At the end of the last eclipse there were a series of papers and presentations given at the January 2011 AAS meeting in Seattle, Washington. The following system diagram of the Epsilon Aurigae Star System is a result of what the current data tells us. It is by no means the final answer. The system is full of surprises and warrants continued study. Photometrically there is a pseudo random magnitude variation that has defied explanation. If appears to be a random pulsation of the star. The variations have a period of about 67 days plus or minus 10 days with an amplitude of around 0.1 magnitudes. Dr. Robert E. Stencel (University of Denver) added that it is predicted that this variation will oscillate more dramatically every one-third of its orbit. Continued spectroscopic observations of the hydrogen alpha line might shed light on these pulsations (Fig. 8.2).

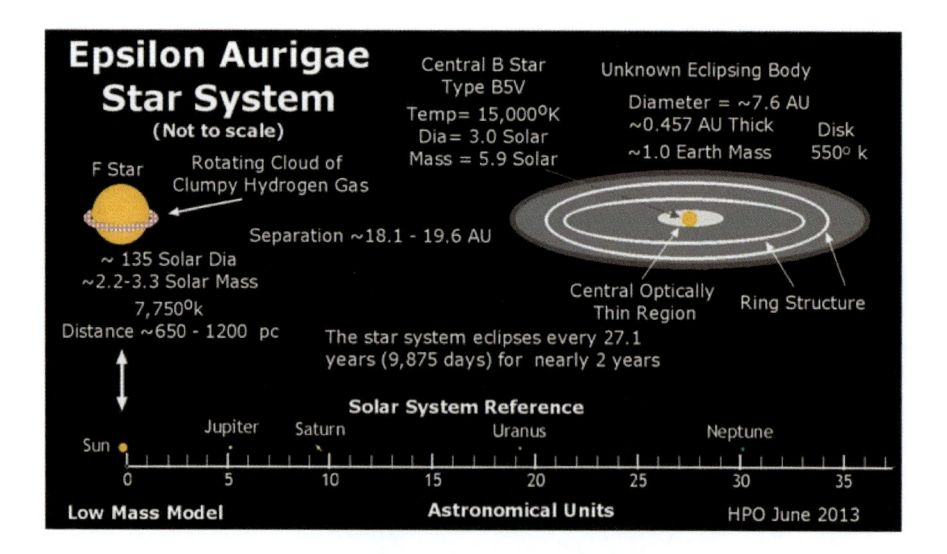

Fig. 8.2 Epsilon Aurigae system diagram

Epsilon Aurigae Spectroscopy

The most fruitful area of the spectrum is the hydrogen alpha region at 6,563 Å. The spectrum in this region is constantly changing and appears to be random. There is a large hydrogen alpha absorption line with blue and red emission horns (sometimes called wings) on either side. The red and blue are not really the colors, but designate the direction in the spectrum, red being the longer wavelength horn and blue being the shorter wavelength horn. Over several days of observations the red and blue horns dance. Sometimes the blue horn is much large than the red and then the reverse happens. Sometimes the horns seem to disappear. It is believed these emission lines are due to a clumpy band of hydrogen gas revolving around the main star (Fig. 8.3).

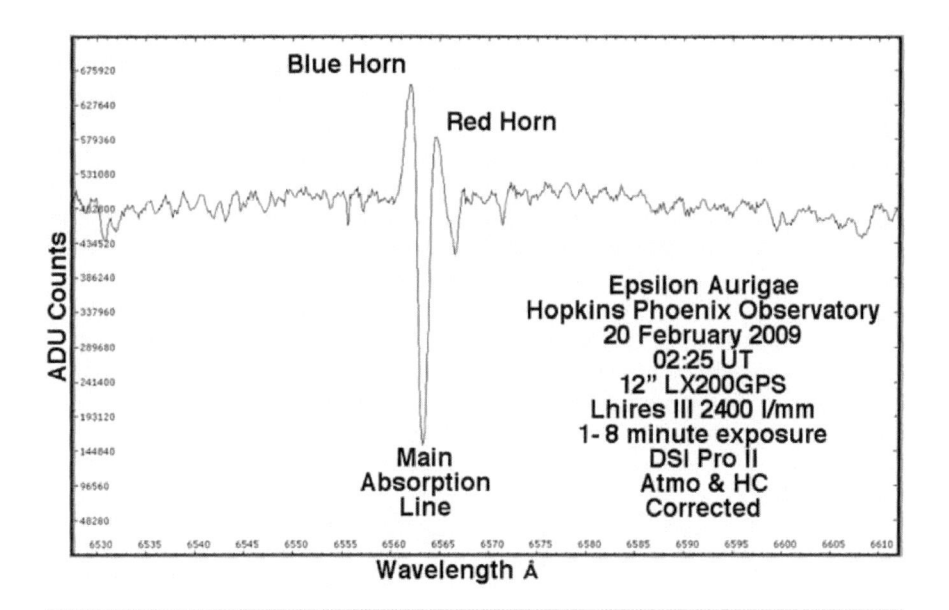

Fig. 8.3 Epsilon Aurigae hydrogen alpha line profile

The sodium D lines are also of interest and easy to image (Fig. 8.4).

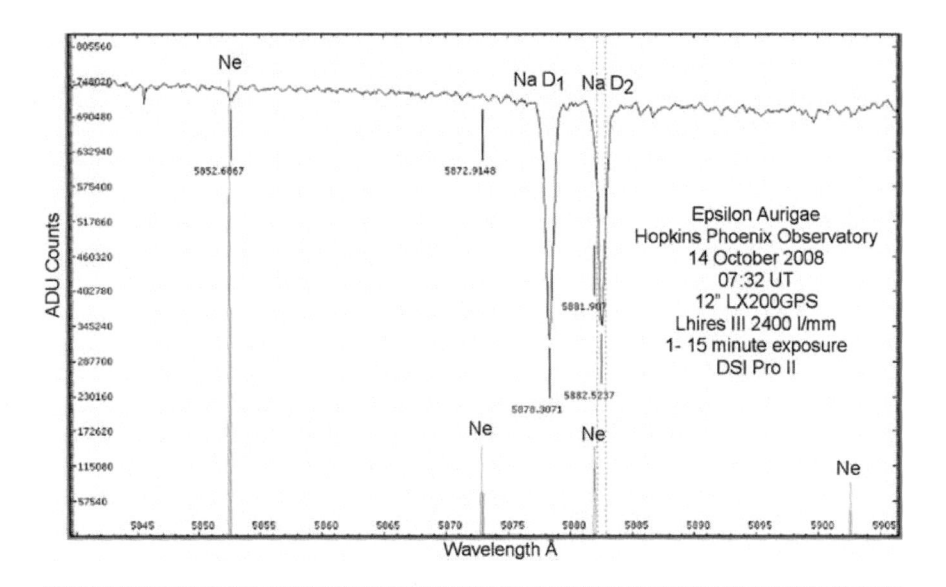

Fig. 8.4 Epsilon Aurigae sodium D line profile

Epsilon Aurigae is an excellent project for someone starting in astronomical spectroscopy. At a declination of +43° and right ascension of 05 h, the system is high in the sky for the Northern Hemisphere during the Fall through Spring seasons. For more information see the International Epsilon Aurigae Eclipse Campaign web site at http://www.hposoft.com/Campaign09.html.

Betelgeuse

Betelgeuse Parameters

Spectrum Type: M2 Iab, V = 0.3 to −1.2
 RA 05 h 56 m 10.3 s, Dec +07d 24′ 25.4″ (2000)

Introduction

Betelgeuse, pronounced "beetle juice," (alpha Orionis) is a very bright variable star, varying between V = 0.3 and −1.2. Variations are not regular and the period varies. This is an old star and is expected to go supernova at any time. It is certainly a star worth continued observations. At the 2012 SAS meeting in Big Bear, California, Dr. john Martin (University of Illinois) suggested this star as a project star for amateur astronomers doing spectroscopy (Fig. 8.5). A web site has been set up as a focal point for continued observations of Betelgeuse and as a source of information sharing. See: http://www.hposoft.com/Betelgeuse/Betelgeuse.html.

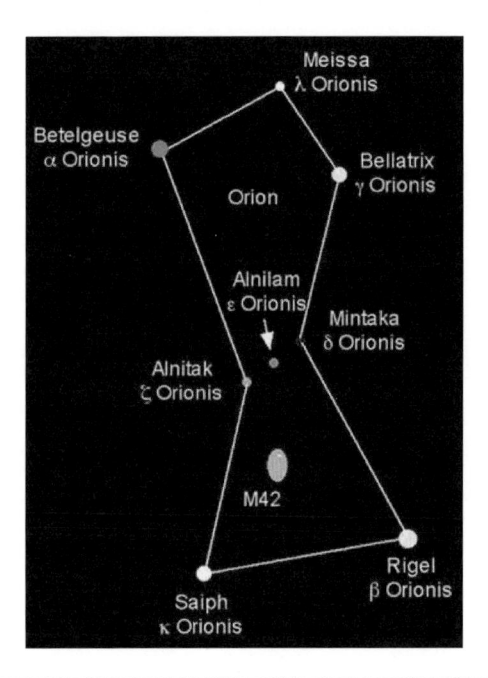

Fig. 8.5 Orion constellation

Because of the brightness of Betelgeuse, most large professional observatories do not monitor it, even though it is constantly changing. This makes it another ideal star for both low and high-resolution amateur spectroscopy.

Betelgeuse Spectroscopy

The Hopkins Phoenix Observatory took the spectrum for the following Figure of a low-resolution spectrum line profile of Betelgeuse. A Star Analyser spectrograph with an 8″ LX 90 SCT was used. Dr. Martin provided the calibration of the lines (Fig. 8.6).

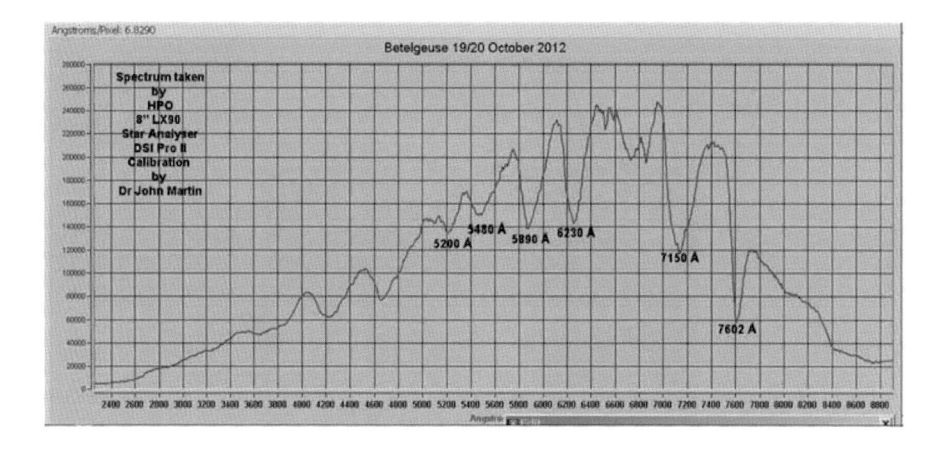

Fig. 8.6 Betelgeuse low-resolution line profile

The Hopkins Phoenix Observatory took the spectrum of the following high-resolution line profile of Betelgeuse using a Lhires III with 600 l/mm grating on a 12″ LX200 GPS SCT (Fig. 8.7).

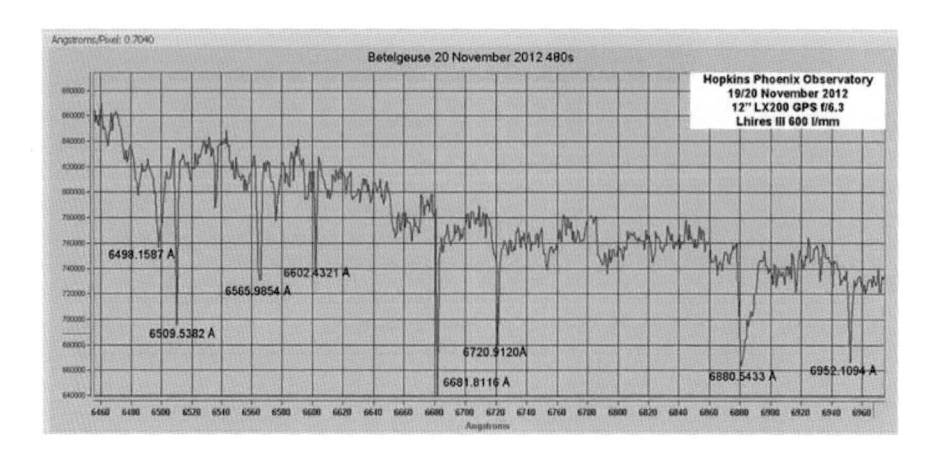

Fig. 8.7 Betelgeuse mid-resolution line profile

Delta Orionis

Delta Orionis Parameters

Spectrum Type: O9.5 II, V = 2.41
RA 05 h 32 m 00.40s, Dec −00d 17′ 56.7″ (2000)

Introduction

Delta Orionis, Mintaka (34 Orionis) is the right star in Orion's belt and is a multiple star system. Dr. Noel Richardson, University of Montreal has suggested continued spectroscopic observations of this star. The hydrogen alpha (6,563 Å) and helium I (6,678 Å) regions are of interest.

Delta Orionis Spectroscopy

The Hopkins Phoenix Observatory took the spectrum for the following Figure of a low-resolution spectrum line profile of delta Orionis. A Star Analyser spectrograph with an 8″ LX 90 SCT was used (Fig. 8.8).

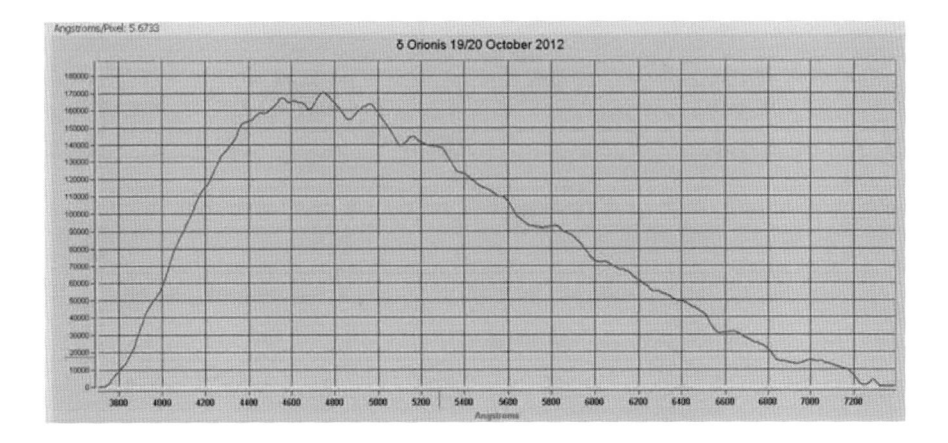

Fig. 8.8 Delta orionis low-resolution line profile

242   8   Astronomical Spectroscopy Projects

The Hopkins Phoenix Observatory took the spectrum of the following high-resolution line profile of delta Orionis using a Lhires III with 600 l/mm grating on a 12″ LX200 GPS SCT (Fig. 8.9).

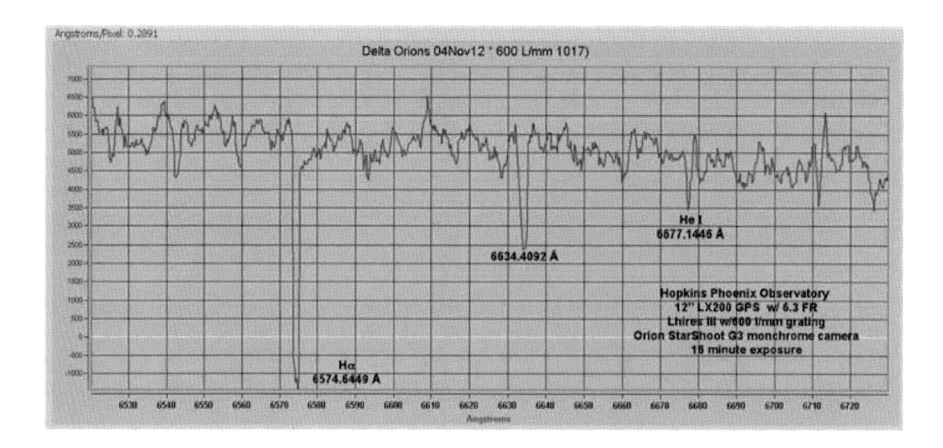

Fig. 8.9 Delta orionis mid-resolution line profile

The Hopkins Phoenix Observatory took the spectrum of the following high-resolution line profile of delta Orionis using a Lhires III with 2,400 l/mm grating on a 12″ LX200 GPS SCT (Fig. 8.10).

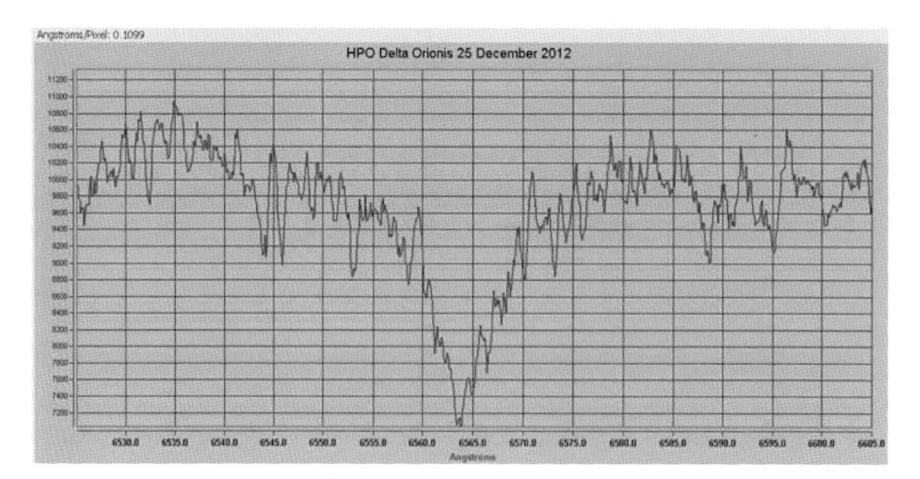

Fig. 8.10 Delta orionis high-resolution line profile

A Campaign with a web site has been set up to follow the star. See http://www.hposoft.com/DOWeb/Campaign.html.

Epsilon Orionis

Epsilon Orionis Parameters

Spectrum Type: B0 Ia, V = 1.70
 RA 05 h 36 m 12.8 s, Dec −01d 12′ 06.9″ (2000)

Introduction

Epsilon Orionis, Alnilam (46 Orionis) is the center star in Orion's belt. Epsilon is a large blue-white supergiant (40 solar masses) star. Because of its simple spectrum epsilon Orionis has been used to study the interstellar medium. It is another star near the end of its life and due to supernova anytime. The helium I absorption line at 6,678 Å is very prominent.

Epsilon Orionis Spectroscopy

The Hopkins Phoenix Observatory took the spectrum of the following high-resolution line profile of epsilon Orionis using a Lhires III with 600 l/mm grating on a 12″ LX200 GPS SCT (Fig. 8.11).

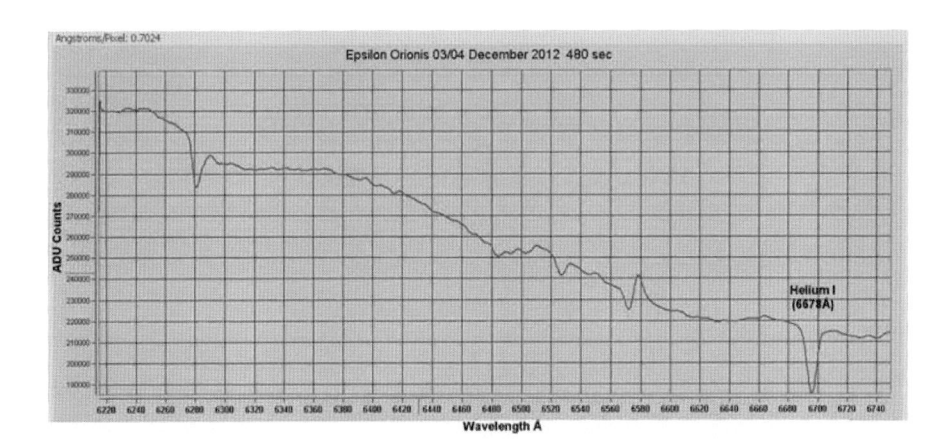

Fig. 8.11 Epsilon orionis mid-resolution line profile

Delta Scorpii

Delta Scorpii Parameters

Spectrum Type: B0, V = 2.31
 RA 16 h 00 m 20.01 s, Dec −22d 37′ 18.14″ (2000)

Introduction

Scorpius is located well to the south, but still within reach of most of the continental United States. Once seen Scorpius is always easy to find again. The curled tail of the scorpion is very obvious (Fig. 8.12).

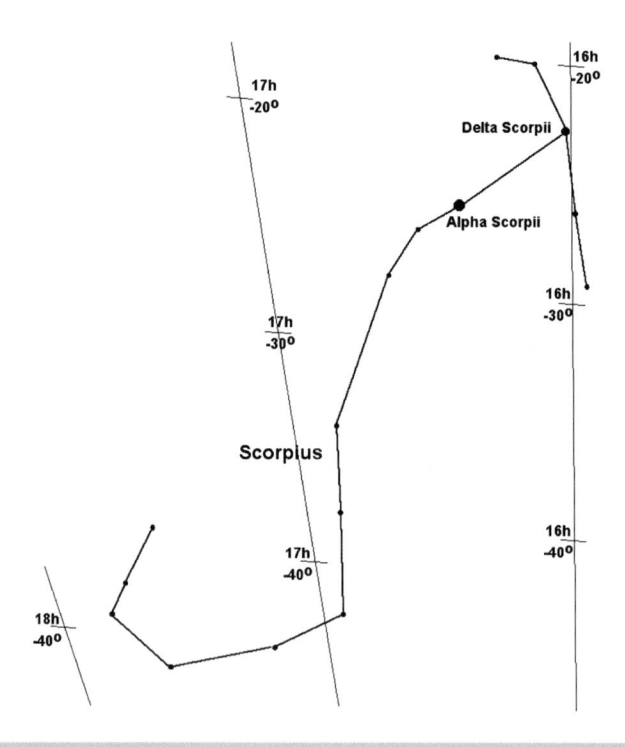

Fig. 8.12 Constellation scorpius

Delta Scorpii is a Be star, but due its popularity it warrants a separate discussion. Delta Scorpii is a bright multiple star eclipsing variable with a period of 10.6 years. There is a B star that orbits the primary every 10 days and a second with the period of 10.6 years. The next eclipse is due in 2022. Of interest are the radial velocity measurements for the star around the time of periastron (time of the nearest point of the stars). The primary star has a mass of 45 Suns and illumination of 38,000 Suns. It is located 470 light years from Earth. Ernst Pollman, Leverkusen, Germany has been instrumental in forming Campaigns to observe this star system. The hydrogen alpha line is the area of most interest. Because of its −22° declination, it provides an excellent target for those in observers from the lower part of the Northern Hemisphere to the mid-Southern Hemisphere. For more information see: http://www.aavso.org/vsots_delsco

P Cygni

P Cygni Parameters

Spectrum Type: B1 Iape, V = 4.81
RA 20 h 17 m 47.2 s, Dec +38d 01′ 58.5″ (2000)

Introduction

On 18 August 1600, Dutch astronomer Willem Janszoon Bleau recorded the existence of a new star in Cygnus, now known as P Cygni. P Cygni is also known as 34 Cyg, HR7763, SAO 69773 and HD193237. P Cygni is a hypergiant blue luminous variable (LBV) star. It has a mass of 30 Suns and a luminosity of 38,000 times that our Sun. Because of its large mass it has burned most of its hydrogen and is near the end of its life. This is another star that could supernova at any time. P Cygni is located about 5,900 light years (1.800 pc) from Earth (Fig. 8.13).

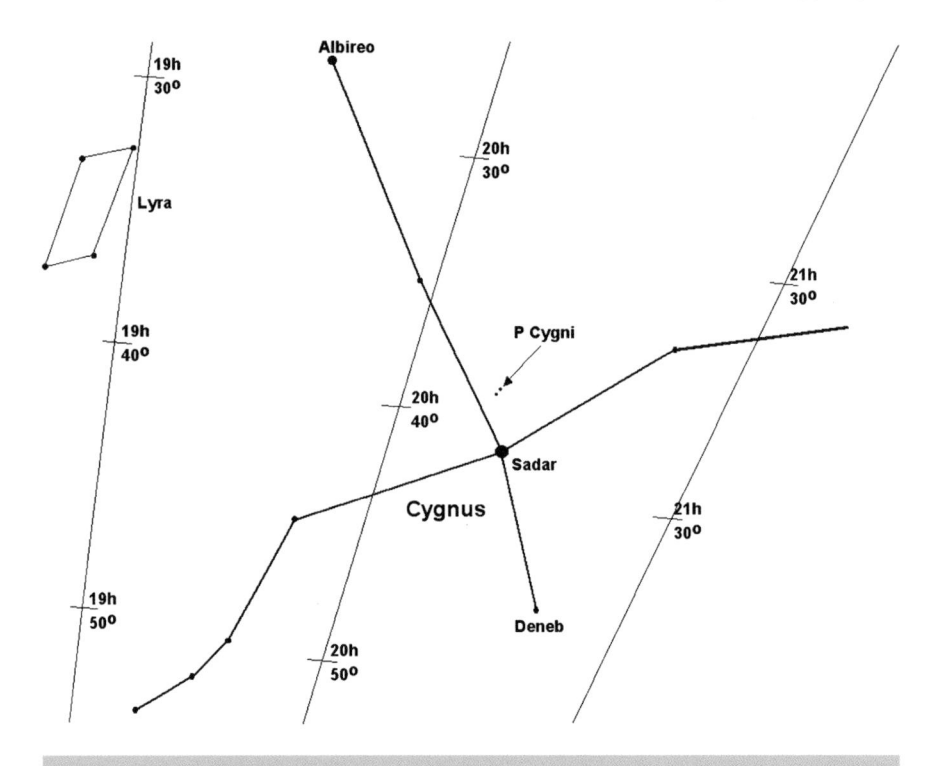

Fig. 8.13 Constellation cygnus

P Cygni Spectroscopy

The hydrogen alpha line is of most interest and has a very large emission line. Ernst Pollmann has been involved in following this star too. The AAVSO has a P Cygni project and while it is mainly photometry observations, spectroscopy is coming into importance. This is an easy and interesting star that is certainly worthy of spectroscopic study (Fig. 8.14). Information can be found at http://www.aavso.org/vsots_pcyg

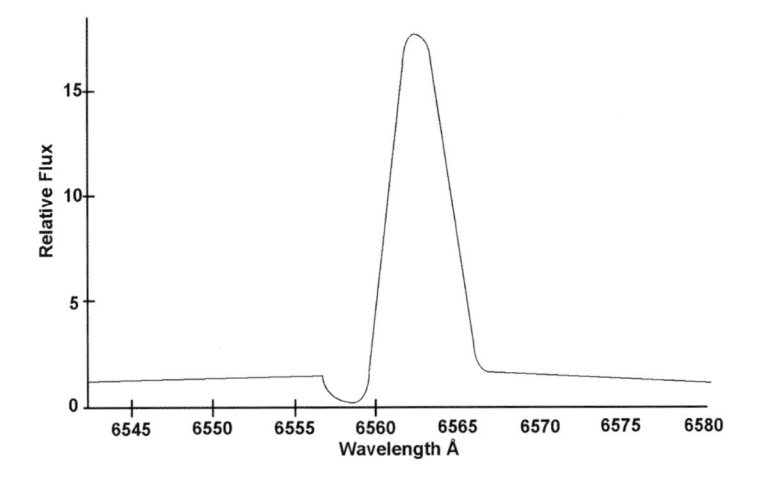

Fig. 8.14 P Cygni typical spectrum line profile

Wolf-Rayet Stars

Introduction

Wolf-Rayet or WR stars are very hot massive (25 solar masses or more) stars at the end of their evolution that unlike most other stars have spectra with emission lines rather than absorption lines (Fig. 8.15). Even the brighter stars will be a challenge

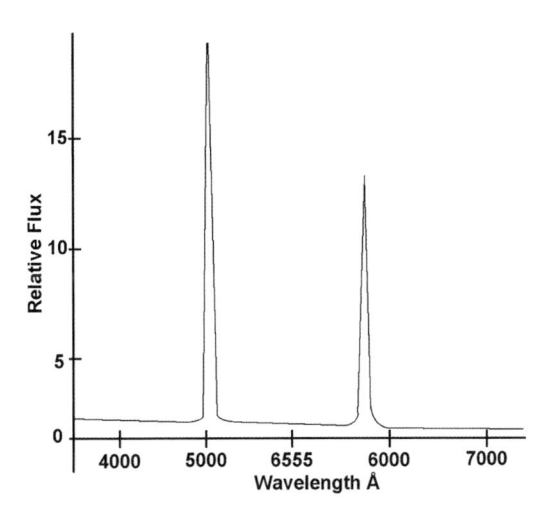

Fig. 8.15 Sample WR star spectrum line profile

for high-resolution spectroscopy, however they should be relatively easy for low-resolution work with a Star Analyser or ALPY 600.

Some Interesting WR Stars

WR 134, HD 191764 in Cygnus, V = 8.07
RA 20 h 10 m 14 s, Dec +36d 10′ 35″ (2000)
WR 135, HD 192103 in Cygnus, V = 8.36
RA 20 h 11 m 53 s, Dec +36d 11′ 50.6″ (2000)
WR 136, HD 192163 in Cygnus, V = 7.48
RA 20 h 12 m 06.5 s, Dec +38d 21′ 17.8″ (2000)
WR 137, HD 191641 in Cygnus, V = 8.15
RA 20 h 10 m 14 s, Dec +36d 10′ 35″ (2000)
WR 140, HD 193793 Sagittarius, V = 13.54
RA 18 h 02 m 04 s, Dec −23d 31′ 41.2″ (2000)

Dr. Noel Richardson (University of Montreal) is coordinating a Campaign to observe WR 134, WR 135 and WR 137. See the following web sites:

http://www.aavso.org/aavso-alertnotice-486
http://www.stsci.de/wr134/index.htm

Be Stars

Introduction

In 1866 Angelo Secchi observed what was to be the first ever star with emission lines, the Be star gamma Cassiopeiae. The "e" comes from "emission" lines in a type B star thus "Be." Be stars are very interesting have quite a following.

Spectroscopy of Be Stars

The hydrogen alpha region shows red and blue horns along with an absorption line similar to epsilon Aurigae. Others show horned emission lines. Many of these stars are bight and an easy challenge for high-resolution spectroscopy. As noted above, delta Scorpii is also a Be star. There are numerous Be stars and there is a group of observers, located mainly in Europe, that have been following these stars for many years. These stars have strong emission lines, particularly the hydrogen alpha line. This contrasts with most other stars that have absorption lines. The hydrogen alpha emission line tends to dance up and down over time and change shape (Fig. 8.16).

Stellar Spectroscopy Projects

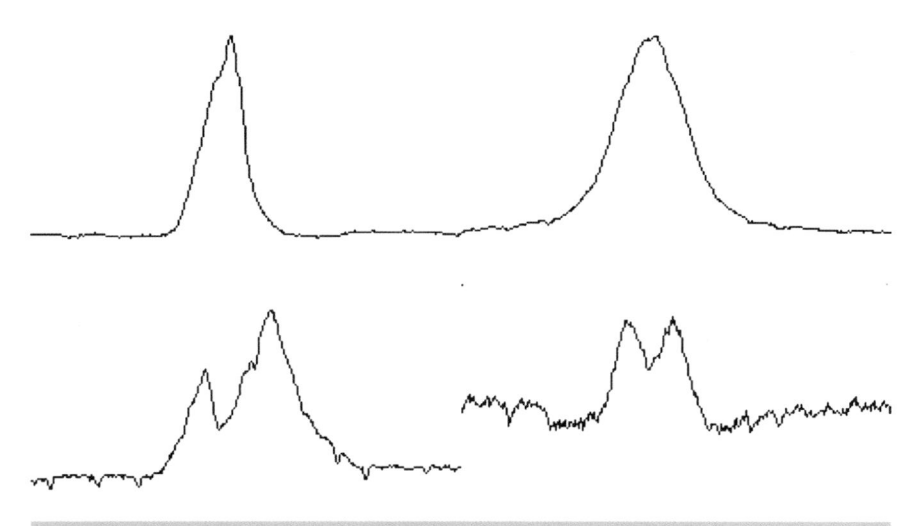

Fig. 8.16 Be star hydrogen alpha emission lines credit: Shelyak

There is a Be Star Atlas on the web site listed below.

Some Popular Be Stars

Beta Lyrae, Shelyak, HR 7106, HD 174638, SAO 67451, V = 3.52
RA: 18 h 50.1 m, Dec: 33d 21′ (2000)
Gammas Cassiopeiae, HR 264, HD 5394, SAO 11482, V = 2.47
RA: 00 h 56.7d, Dec: 60d 43′ (2000)
Zeta Tauri, HR 1910, HD 37202, SAO 77336, V = 3.00
RA: 05 h 37.6, Dec: 21d 09′ (2000)
Omega Orionis, HR 1934, HD 37490, SAO 113001, V = 4.57
RA: 05 h 39.2, Dec: 04d 07′ (2000)
SS Leporis, HR 2148, HD 41511, SAO 151093, V = 4.93
RA: 06 h 05, Dec: −16d 29′ (2000)
HD 206773(1), SAO 33677, V = 6.87
RA: 21 h 42.4, Dec: 57d 44′ (2000)

For more information see: http://basebe.obspm.fr/basebe/

Quasar Projects

Introduction

Quasars are too faint for high-resolution work with modest sized telescopes, but offer an easy challenge for low-resolution spectroscopy with spectrographs like the Star Analyser and ALPY 600. Being able to measure the red shift of a quasar is exciting when one remembers originally the 200″ Hale telescope was needed for even the brightest Quasar 3C 273. Quasar 3C 273 is the quasar most observed by amateurs. Other Quasars are considerably fainter and present a much greater challenge. The following are two that may be of interest to amateurs. Quasar 3C 48 is much fainter than 3C 273, but with a modest telescope and a low-resolution spectrograph it may be possible to image a spectrum of it with a long time exposure.

Some Interesting Quasars

3C 273 Located in the Constellation Virgo, V Magnitude 12.9
RA 12 h 29 m 06.7 s, Dec +02d 03′ 09″ (2000)
3C 48 Located in the Constellation Triangulum, V Magnitude 16.2
RA 01 h 37 m 41.1, Dec +33d 09′ 32″ (2000)

Supernova Projects

Introduction

Since supernovae are not predictable, one has to be diligent to watch for announcements of new suspected supernovae. While most supernovae are too faint for high-resolution work with modest telescopes and reasonable exposure times, low-resolution offers some interesting possibilities. The AAVSO has a Nova Project forum. You do not need to be a member to subscribe to the forum.

See the AAVSO web site at http://www.aavso.org.

While most of the AAVSO nova and supernova activity is visual magnitude estimations and photometry, there is an opening for spectroscopy work too. It may be possible to gain some fame by being the first to get a spectrum of a new supernova and classify it. For more information on supernovae work using low-resolution spectroscopy see Chap. 1 Supernovae Classification. For notifications of new supernovae see the following:

Supernova Contacts

To report astronomical discoveries, see: http://www.cbat.eps.harvard.edu/HowToReportDiscovery.html

For information on recent supernovae, see: Rochester Astronomy Supernova Web Site at http://www.rochesterastronomy.org/supernova.html

IAU Central Bureau of Astronomical Telegrams http://www.cbat.eps.harvard.edu/cbet/RecentCBETs.html

The Astronomer's Telegram http://www.astronomerstelegram.org/

The CBA Belgium Automated Supernova-Search Program http://tonnyvanmunster.ipage.com/supernovae.htm

Appendix A

Astronomical Time

Introduction

To do serious astronomy one must have a good understanding of time.

Time is simple, right? Actually time is not simple at all, we are just used to it. Take for example there is no zero on clocks. Sure there is a ten which is a 10, but that is ten, not zero. Older clocks use the Roman Numeral X for ten. There is no Roman numeral for zero. Now some may say zero is 00:00:00 h which is midnight. Indeed, some modern time uses a 24 h system and it starts at midnight at 00:00:00 h. Military or Zulu time is like this and in astronomy, Universal Time (UT). All these 24 h times are based on the time at that instant on the Greenwich Meridian (GMT) and start at midnight equal to 00:00:00 h. Some digital clocks allow local time to also be 24 h based and not based on GMT, but local midnight. When I say 00:00:00 h I mean **hh:mm:ss**, where **hh** are two digits for the hour (06), **mm** are two digits for minutes (13) and **ss** for seconds (33.874564). You can make the time as accurate and precise as you wish by adding numbers to the right of the decimal point in the second's part. Time is now 06:13:33.874564 (06 h, 13 min and 33.874564) seconds).

Astronomical Times	Astronomical Dates
Local Standard Time	Decimal Date
Local Daylight Savings Time	Julian Date (JD)
Universal Time (UT)	Modified Julian Date (MJD)
Sidereal Time	Reduced Julian Date (RJD)
Local Sidereal Time	Heliocentric Julian Date (HJD)

Astronomical Time and Date

Astronomical Time

Local Standard Time

This is the time in a particular time zone. Time zones are divided up for convenience and sometimes very confusing. Slightly North, South, East or West and it could be a different hour.

In the continental United States we have 4 time zones Eastern Standard Time (EST), Central Standard Time (CST), Mountain Standard Time (MST) and Pacific Standard Time (PST). There are similar time zones around the world. Based on UT, EST is 5 h behind. When UT is 23:00:00 h EST is (−5 h) 18:00:00 h, CST is (−6 h) 17:00:00, MST is (−7 h) 16:00:00 and PST is (−8 h) 15:00:00.

Things get confusing when say EST is 20:00:00 as it is 01:00:00 h UT the next day. In astronomy it is always good to supply a double date where the first date is the evening and the second is the next morning's date. For an observation made on the evening of 29 December 2009 at 18:00:00 h EST the reported date and time would be 29/30 December 2009 23:00:00 UT. Both the UT date and Local date would be 29 December.

An observation on the evening of 29 December 2009 at 20:00:00 h EST would be reported as 29/30 December 2009 01:00:00 UT. The UT date would be 30 December, but the Local date 29 December.

Local Daylight Savings Time

In the continental USA the time zones then become EDT (UT−4), CDT (UT−5), MDT (UT−6) and PDT (UT−7). While this may seem simple, it is not. Not everywhere uses Daylight Time. In fact most, but not all, of Arizona never uses Daylight Time so during the months daylight time is in effect, most of Arizona is still on Standard Time (MST) and thus PDT is the same as MST. With proper adjustments all is well. Note: Universal Time does not change with Daylight Savings Time.

Universal Time (UT)

Also Known a UTC or Coordinated Universal Time

In astronomy, most of the time people observe events outside of the Earth. It behooves us to find a standard whereby astronomical events can use the same time. This is where Universal Time or UT comes in. It is the same as Greenwich Mean Time defined as 12:00:00 h when the Sun transits the zero longitude line that passes through Greenwich, England. While this can vary by up to 16 min (due to the equation of time or variation of the Earth's elliptic speed and axial tilt), UT is now set by an atomic clock (UTC). Generally UTC and UT are the same time.

Appendix A

Sidereal Time

Lines of latitude going north and south of the Equator and lines of Longitude going east and west of the GMT meridian divide the Earth up. From our perspective on Earth, the sky can be thought of as a sphere and divided up similarly.

Instead of lines of Latitude and Longitude, the celestial sphere has lines of Declination or Dec (latitude) and Right Ascension or RA (Longitude). The celestial Equator is above the Earth's Equator. As the Earth rotates on its axis the celestial sphere appears to turn from East to West. The Declination is just like Latitude going from zero degrees at the Equation to +90° at the North Pole and −90° at the South Pole. Longitude or Right Ascension is divided up into 24 h instead of degrees.

Because of the Earth's motion around the Sun a sidereal day is slightly less than 24 h. A sidereal day is 23 h 56 min and 04.091 s. The zero sidereal hour is a meridian on the celestial sphere where the Sun crosses the celestial Equator in March. This is also known as the first point of Aries and Vernal Equinox. The RA of a Star is its sidereal time.

When that RA time of a star is the same as the local sidereal time, that star is on your meridian. The local time shifts about 2 h a month for the sidereal time. In other words a star is on your meridian at 11:00:00 h UT (04:00:00 h MST) on 15 September. The star will be on the meridian at 09:00:00 h UT (02:00:00) MST) on 15 October and 07:00:00 h UT (00:00:00 h MST) in November.

Local Sidereal Time

Local Sidereal Time is just the RA that is on your meridian at a given time. Epsilon Aurigae has a RA of 05 h, 02 min. When the star is on your meridian, your Local Sidereal Time is 05:02.

Astronomical Date

Decimal Date

Many applications require the date to be in decimal form. What that means is the day of month followed by a period and then a number representing the percent of the day.

$$\text{For example}: 12.50000$$

This would be the 12th day of the month at noon. It does not specify if it is local or UT. That is determined by the application requirement.

$$\text{A more complex time is}: 19.09444 \text{ UT}$$

This for 18 February 2012 at 19:18 MST which relates to the next day for Universal Time 19 February 2012 and a time of 02:16 which converts to the: 19.09444 UT. In this case the month and year are not specified.

Julian Date uses a similar decimal technique.

Julian Date

Keeping track of dates over long periods when there are leap years involved and even changes in calendars can be a big problem. This is especially true for astronomy when you may want to go back in time hundreds of years to an event and know how much time there was between then and now. To make this task easier, a system using Julian Date was devised.

Julian Date is the number of days that have elapsed since noon on 1 January 4713 B.C. and is given in the form of decimal days, not in hours, minutes, and seconds. The Julian Day begins at noon Greenwich time or 12 h UT. Leap years and calendar changes do not matter.

Examples of Julian Dates are:

$$11 \text{ January } 1988 \text{ at } 2^h 45^m \text{ UT,}$$
$$JD = 2,447,171.6146$$

$$11 \text{ January } 1988 \text{ at } 12^h 2^m \text{ UT,}$$
$$JD = 2,447,172.0333$$

What is the origin of the Julian Date system? Contrary to some beliefs, Julian Date has no connection with the Julian calendar and was not named after Julius Caesar. Instead Joseph Justus Scaliger, in 1583, developed the Julian Period. He multiplied the lengths of three cycles: the 28-year solar cycle, the 19-year lunar cycle, and the 15-year cycle of the Roman Indiction (used in calculating the date of Easter). The resulting period ($28 \times 19 \times 15$) is 7980 years, which passed through zero in the year 4713 BC. This is a very convenient date because all recorded history, including documented astronomical events, has occurred after this date. Astronomers adopt Julian Dates because the time interval between events is independent of the day of the week, month, or year.

If a table is not available, such as those in the Astronomical Almanac, then Julian Date can be calculated using the formula

$$JD = 367 * Y - \text{Int}\,(7 * (Y + \text{Int}\,((M + 9)/12)))/4) + \text{Int}\,(275 * M / 9)$$
$$+ D + 1721013.5 + UT / 24$$

where

Y, M, D, are the year, month, day of month and UT is Universal Time (in 24-h, decimal format).

The function **int** is the integral part of the quotient resulting from the division of two integers (i.e., Int(3/4)=0 and Int(5/4)=1). The formula for JD is valid for the years 1901–2099.

Example: What is the Julian Date corresponding to 4 h 25 m 16 s UT on September 9, 1990?

$$Y = 1990 \quad M = 9 \quad D = 9 \quad UT = 4.42111^h$$

$$JD = 367 \times 1990 - \text{Int}(7(1990 + \text{Int}(18/12))/4) + \text{Int}(275 \times 9/9) + 9 + 1721013.5 + 4.42111/24$$

$$JD = 730330 - 3484 + 275 + 9 + 1721013.5 + 0.18421$$

$$JD = 2,448,143.68421$$

If time is recorded to the nearest minute or second, the JD can be calculated to the nearest 0.001 or 0.00001 day, respectively.

An even simpler method of determining JD is to know the JD on the first day of the month and just add the month's day to that date.

Modified Julian Date (MJD)

Because space is sometimes limited, particularly on plots, it is advisable to use a reduced or modified Julian Date. The MJD is some part of the JD, usually the last 4 or 5 digits plus a 0.5 day.

$$MJD = 55,096.4965$$

where

$$JD = MJD + 2,400,000.5$$

Reduced Julian Date (RJD)

There is some debate as to what Modified Julian Date (MJD) and Reduced Julian Date (RJD) mean. The purpose is to provide a focus on the more important and changing numbers. We prefer the use of Reduced Julian Date (RJD=JD−2,450,000 or depending on the time frame involved RJD=JD−2,400,000).

JD	RJD
2,455,127.13	5,127.13
2,455,128.12	5,128.12
2,455,129.15	5,129.15
JD=2,455,000+RJD	

or

2,455,129	55,129
2,455,458	55,458
2,455,691	55,691
JD = 2,450,000 + RJD	

Heliocentric Julian Date

Because the speed of light is finite and the earth is traveling around the sun with an orbital diameter of nearly 3×10^8 km, it is both convenient and necessary to adopt a time system that removes the effects of the earth's motion. In a geocentric reference frame, the light from a star in the general direction of the sun (i.e., near conjunction) requires approximately 16 additional minutes to reach the earth compared to a star at opposition. By using the center of the sun as reference point for all time measurements (i.e., a heliocentric system) the errors due to light-travel time across the earth's orbit are eliminated. For observations where timing to the minute or second is important, it is essential to apply this heliocentric correction. For astronomical photometry, precise timing information is specified in terms of Heliocentric Julian Date, or HJD. The correction to HJD can be found using the following equations:

$$HJD = JD_{Hel} = JD_{Geo} + \Delta t$$

$$\Delta t = -T R \left(\cos\lambda \cos\alpha \cos\delta + \sin\lambda (\sin\varepsilon \sin\delta + \cos\varepsilon \cos\delta \sin\alpha)\right)$$

where

T = Light – travel time for 1 A.U.
$(= 499 \text{ sec or } 0.0057755 \text{ days})$

R = Earth – Sun distance in A.U.

λ = Longitude of the sun

α = Right Ascension of the star

δ = Declination of the star

ε = Obliquity of the ecliptic $(= 23.45 \text{ degrees})$

Appropriate values for R and λ can be found in the Astronomical Almanac for each observing night. Because R and λ change rather slowly, a unique value of Δt can be calculated for each star and applied to all observations of that star on a given night. There are alternative expressions for Δt that use the rectangular coordinates of the sun, X and Y, rather than R and λ.

Appendix A

Julian Date List

Julian Dates for the first of each Month from 1 January 2008 to 1 December 2013. Remember, the Julian Day begins at noon Greenwich time or 12 h UT.

1 January 2011	**JD = 2,455,563**
1 February 2011	JD = 2,455,594
1 March 2011	JD = 2,455,622
1 April 2011	JD = 2,455,653
1 May 2011	JD = 2,455,683
1 June 2011	JD = 2,455,714
1 July 2011	JD = 2,455,744
1 August 2011	JD = 2,455,775
1 September 2011	JD = 2,455,806
1 October 2011	JD = 2,455,836
1 November 2011	JD = 2,455,867
1 December 2011	JD = 2,455,897
1 January 2012	**JD = 2,455,928**
1 February 2012	JD = 2,455,959
1 March 2012	JD = 2,455,988
1 April 2012	JD = 2,456,019
1 May 2012	JD = 2,456,049
1 June 2012	JD = 2,456,080
1 July 2012	JD = 2,456,110
1 August 2012	JD = 2,456,141
1 September 2012	JD = 2,456,172
1 October 2012	JD = 2,456,202
1 November 2012	JD = 2,456,233
1 December 2012	JD = 2,456,263
1 January 2013	**JD = 2,456,294**
1 February 2013	JD = 2,456,325
1 March 2013	JD = 2,456,353
1 April 2013	JD = 2,456,384
1 May 2013	JD = 2,456,414
1 June 2013	JD = 2,456,445
1 July 2013	JD = 2,456,475
1 August 2013	JD = 2,456,506
1 September 2013	JD = 2,456,537
1 October 2013	JD = 2,456,567
1 November 2013	JD = 2,456,598
1 December 2013	JD = 2,456,628

(continued)

(continued)

1 January 2014	**JD = 2,456,659**
1 February 2014	JD = 2,456,690
1 March 2014	JD = 2,456,718
1 April 2014	JD = 2,456,749
1 May 2014	JD = 2,456,779
1 June 2014	JD = 2,456,810
1 July 2014	JD = 2,456,840
1 August 2014	JD = 2,456,871
1 September 2014	JD = 2,456,902
1 October 2014	JD = 2,456,932
1 November 2014	JD = 2,456,963
1 December 2014	JD = 2,456,993
1 January 2015	**JD = 2,457,024**
1 February 2015	JD = 2,457,055
1 March 2015	JD = 2,457,083
1 April 2015	JD = 2,457,114
1 May 2015	JD = 2,457,144
1 June 2015	JD = 2,457,175
1 July 2015	JD = 2,457,205
1 August 2015	JD = 2,457,236
1 September 2015	JD = 2,457,267
1 October 2015	JD = 2,457,297
1 November 2015	JD = 2,457,328
1 December 2015	JD = 2,457,358
1 January 2016	**JD = 2,457,389**
1 February 2016	JD = 2,457,420
1 March 2016	JD = 2,457,449
1 April 2016	JD = 2,457,480
1 May 2016	JD = 2,457,510
1 June 2016	JD = 2,457,541
1 July 2016	JD = 2,457,571
1 August 2016	JD = 2,457,602
1 September 2016	JD = 2,457,633
1 October 2016	JD = 2,457,663
1 November 2016	JD = 2,457,694
1 December 2016	JD = 2,457,724

(continued)

(continued)

1 January 2017	**JD = 2,457,755**
1 February 2016	JD = 2,457,786
1 March 2016	JD = 2,457,814
1 April 2016	JD = 2,457,845
1 May 2016	JD = 2,457,875
1 June 2016	JD = 2,457,906
1 July 2016	JD = 2,457,936
1 August 2016	JD = 2,457,967
1 September 2016	JD = 2,457,998
1 October 2016	JD = 2,458,028
1 November 2016	JD = 2,458,059
1 December 2016	JD = 2,458,089
1 January 2018	**JD = 2,458,120**
1 February 2016	JD = 2,458,151
1 March 2016	JD = 2,458,179
1 April 2016	JD = 2,458,210
1 May 2016	JD = 2,458,240
1 June 2016	JD = 2,458,271
1 July 2016	JD = 2,458,301
1 August 2016	JD = 2,458,332
1 September 2016	JD = 2,458,363
1 October 2016	JD = 2,458,393
1 November 2016	JD = 2,458,424
1 December 2016	JD = 2,458,454

Appendix B

FITS Header

Introduction

Images that are saved as a .fit or .fits (**F**lexible **I**mage **T**ransport **S**ystem) file contain header information regarding the image. Different software applications add different information. This information is added automatically as text to the top of the Image file.

Creating a FITS Header

While the format is standard, the contents of a FITS header can vary from program to program. The software program used to control the CCD camera and save the image usually has the capability of saving the image as a FITS file, along with other image formats. When a FITS header is created, the software program adds information to the header. Information, such as the date, time and length of the exposure is added automatically. This is a good reason to make sure your computer clock is set correctly. You can even set your computer clock for Universal Time to make the data even better. Some software allows you to enter information one time that will be added to all the headers. This can be your name, observatory, information about the telescope and other equipment, location. You can add files of information by using all upper case with no spaces followed by an equal sign and then the information.

For example new fields created called **SPECTROGRAPH** and **GRATING**:

SPECTROGRAPH=Lhires III
GRATING=2,400 l/mm grating

You can make as many of those additions as you wish. Unless the imaging software lets you add the new fields you will need to use a text editor and add them to each file after the image file has been created. If the image software adds them then it will be done automatically. Most imaging software will add the date, time, exposure time, number of stacked images, observer, observatory, telescope and other fixed information. Again, for this to be automatically added it must be setup with the software prior to taking the images. Be sure to set the computer's date and clock. It may be helpful to set it to Universal Time instead of local time.

This does not work with other formats, e.g., .jpeg, .gif, .png or .bmp.

Reading a FITS Header

The FITS header can be read with any text or word processing program. The basic format is standard, but as noted above, the information varies with different imaging software and can have additional data fields added via text program.

Unless otherwise noted, the time reported in the header is the time at the start of the observation. It is best to check your software to see what time is used. Some programs use a mid-time and others report both the start and mid-exposure. For log information the mid-observation time is a more accurate time. The mid-exposure time can be determined manually (observation start time plus one half the exposure time) if not supplied in the header. Use the mid-time for the logged observation time. For exposures of a few seconds, this will not be of concern, but is important for long time exposures in minutes.

The following is what Envisage enters for the FITS Header:

```
SIMPLE= T/file does conform to FITS standard
BITPIX=-32 / number of bits per data pixel
NAXIS= 2 / number of data axes
NAXIS1= 748 / length of data axis 1
NAXIS2= 577 / length of data axis 2
EXTEND= T/FITS dataset may contain extensions
COMMENT FITS (Flexible Image Transport System) format is
defined in 'Astronomy
COMMENT and Astrophysics', volume 376, page 359; bibcode:
2001A&A...376..359H
EXPTIME= 0.5 / Total Exposure Time
SNAPSHOT= 1 / Number of stacked images
FOCALLEN= 1048 / Focal Length
OBSERVER='Jeff Hopkins' / Observer Name
APTDIA= 305 / Aperture Diameter in mm
```

```
TELESCOP= 'LX200 GPS' / Telescope Model
MOSAIC = NONE / Mosaic Pattern
CCD-TEMP= 17.5 / CCD temp in C
OBJECT= Eaur05sec24Jan12'/ Name of object
INSTRUME='DSI Pro-DSI-2'/Imager Model and Name
USERSETU='Spectroscopy Lhires III 2400 l/mm'/Additional
user setup data
TIME-OBS= '01:40:42' / UTC Time
DATE-OBS= '2012-01-25'/ UTC Date
BSCALE= 1. /
BZERO=0. /
BACKGRND= 5845 /
RANGE= 33516 /
END
```

The Orion StarShoot software does not have a user settings input, but provides the following header:

```
SIMPLE= T / C# FITS: 1/17/2012 19:19:31
BITPIX= -32
NAXIS=2 / Dimensionality
NAXIS1= 752
NAXIS2=582
INSTRUME='Orion G3 Deep Sky Monochromatic Camera'/Camera
used
SWCREATE= 'Orion Camera Studio'/Name of the software
that created the image
IMAGTYP = 'Dark Frame' / Type of image
CCD-TEMP= 4.60820512820513/CCD temperature at start of
exposure in C
ENCODING= 'RAW / Encoding type: RAW/RGB/YcbCr/Multichannel
PEDESTAL= 0 / Pedestal
SET-TEMP=4.95589743589743 / CCD temperature setpoint in C
YBINNING= 1 / Binning factor in height
EXPTIME= 480 / Exposure time in seconds
DATE_OBS= '2012-01-17 T07:19:31'/Date
EXTEND= T / Extensions are permitted
EXPOSURE=480 / Exposure time in seconds
XBINNING= 1 / Binning factor in width
YPIXSZ= 8.30000019073486/Pixel Height in microns (after
binning)
XPIXSZ= 8.60000038146973/Pixel Width in microns (after
binning)
END
```

Appendix C

Important Wavelengths

Na D lines
  D1 – 5,889.950 Å
  D2 – 5,895.929 Å

Hydrogen Balmer Lines
  Hydrogen alpha (Hα) – 6,562.281 Å
    Is actually a doublet at 6,562.72 and 6,562.85 Å
    The wavelength stated is a weighted average.
  Hydrogen beta (Hβ) – 4,861.35 Å
  Hydrogen gamma (Hγ) – 4,340.472 Å
  Hydrogen delta (Hδ) – 4,101.734 Å
  Hydrogen epsilon (Hε) – 3,970.075 Å
  The following lines will not be seen without special equipment.
  Hydrogen zeta (Hζ) – 3,889.054 Å
  Hydrogen eta (Hη) – 3,835.397

Helium I – 6,678.15 Å

Neon Lines around the Hydrogen Alpha Line
  6,506.53 Å
  6,532.88 Å
  6,598.95 Å
  6,678.28 Å

Neon Lines around the Sodium D Lines
  5,852.49 Å
  5,872.83 Å
  5,881.90 Å

Laser Wavelengths
   He/Ne Laser – 6,330 Å
   Red Pointer a – 6,621 Å
   Red Pointer b – 6,571 Å
   Red Pointer c – 6,560 Å
   Green Pointer – 5,320 Å
   Violet Pointer – 4,050 Å

Telluric (H_2O) Spectral Line Wavelengths Around the Hydrogen Alpha Line
   6,532.359 Å
   6,534.000 Å
   6,536.720 Å
   6,542.313 Å
   6,543.907 Å
   6,547.705 Å
   6,548.622 Å
   6,552.629 Å
   6,553.785 Å
   6,557.171 Å
   6,560.499 Å
   6,564.196 Å
   6,568.806 Å
   6,572.072 Å
   6,574.847 Å
   6,580.786 Å
   6,586.596 Å
   6,599.324 Å

Glossary

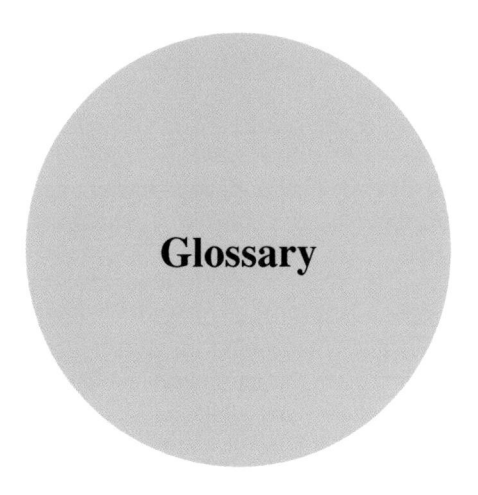

4 – Shooter The first astronomical digital CCD camera designed and built by James Gunn and used on the Hale 200″ telescope.

Absolute Calibration The calibration of the Y-axis (intensity) of a line profile to indicate the flux versus wavelength.

Absorption Spectrum Absorption spectrum are spectra with the presence of dark portions in a bright continuum spectrum due to absorption of photons by elements present in the source that absorb specific wavelengths.

ADC Analog to Digital Converter.

ADU Analog to Digital Unit. These are counts that relate to the number of photoelectrons in a CCD pixel.

Air Mass An approximation for fractional column of additional air along the line of sight, relative to straight up (zenith) which is defined as one air mass. Air mass varies approximately with the inverse cosine (90° minus elevation angle).

AIP4WIN This is an image processing software program that works on Windows. It has no drivers for any CCD cameras. Cameras are controlled with other software.

Aldebaran A bright star also known as alpha Tau. α Tau is a star with an interesting spectrum.

Alpha Lyrae A star also known as Vega.

Alpha Orionis Also known as Betelgeuse or α Ori.

Alpha Tau Also known as Aldebaran or α Tauri.

ALT/AZ Altitude/Azimuth. A motion up and down and rotated.

Ångströms (Å) An Ångström is a measure of light's wavelength. One nm = 10 Å.

Arc second One arc second is an angular unit which is one 60th of an arc minute, which in turn is one 60th of one degree. The Sun and Moon subtend about 30 arc minutes as seen from earth.

ATIK A CCD camera suitable for astronomical use.

Atmospheric Lines Atmospheric lines are spectral absorption lines due to photon absorption of oxygen, water and carbon dioxide in the Earth's atmosphere. Also known as telluric lines.

ASCII American Standard Code for Information Interchange.

AU Astronomical Unit, the average earth-sun separation. One AU = 93 million miles or 150 million km.

AutoStar AutoStar or more correctly AutoStar Suite. This is a suite of software by Meade to display a planetarium program, provide CCD camera control and image acquisition, provide telescope control for GOTO and tracking options. The image processing part processes images and can extract net ADU.

Barycenter Barycenter is a number describing a pixel position determined by two bracketing points. By using interpolation, sub-pixel positions can be determined. This is also the centroid of the object or center of gravity.

Balmer Series The Balmer Series is a sequence of spectral lines formed when electrons in hydrogen atoms transition between level 2 and higher levels – visible part of the spectrum.

Bellatrix A bright star also known as alpha Orionis.

Be Stars Type B stars that have strong emission lines, particularly the hydrogen alpha line.

Beta Lyrae Also known as Shelyak, is a famous bright eclipsing (period of 12.9414 days) binary star system with an interesting spectrum.

Beta Orionis A bright star also known as Bellatrix and β Ori.

Betelgeuse A bright star also known as alpha Orionis.

Binning Binning is the averaging of two or more adjacent pixel ADU values.

Black Body A black body is an object that absorbs all light that falls on it. No electromagnetic radiation passes through it and none is reflected. Because no light is reflected or transmitted, the object appears black when it is cold.

Blazed Grating A blazed grating is a diffraction grating that has lines slanted from the perpendicular at angles that optimize certain wavelengths.

Brackett Series The Brackett Series is a sequence of spectral lines formed when electrons in hydrogen atoms transition between level 4 and higher levels – near infrared part of the spectrum.

CCD A CCD is a Charge-Coupled Device, a silicon based array detector found in many digital cameras.

Centroid Center of gravity or barycenter.

Cepheid A class of intrinsic variable stars used to determine stellar distances using the inverse-square law. The star's Absolute Magnitude is a function of the period of pulsation.

CFL Compact Fluorescent Light.

Charge-Bubble Device Original name at AT&T Bell Labs for the CCD. Used for computer memory applications.

CMOS Complimentary Metal Oxide Semiconductor.

Collisional Broadening Collisonal Broadening is the effect of electron pressure/motion on spectral line width.

Continuous Spectrum A continuous spectrum is the color pattern predicted by Planck's radiation law. It is a spectrum with no absorption or emission lines.

Continuum This is the spectrum of a star created by the star's internal nuclear reactions before and absorption due to the star's atmosphere or other absorption.

Dark Frame This is a digital camera's image obtained when no light is allowed to reach the detector. It is useful in calibrating the background contribution in the images.

Data Acquisition This is the acquiring of star data such as by taking an image of the star's spectrum.

Data Reduction This is the improvement of accuracy of a raw photometric magnitude by applying various calibration functions thereby reducing a raw magnitude to an extraterrestrial magnitude.

Decimal Date A date where the time of day is added in decimal form. For example noon of on the 15th of the month would be 15.50.

Diffraction Grating This is a surface with ruled lines used to produce a spectrum. There are two types, transmission and reflection.

Dispersion How spread out a spectrum is determines its dispersion in Å/pixel. Dispersion is usually related inversely to resolution.

Doppler Effect The shifting of frequency/wavelength of light as it approaches or departs an observer. The same effect is true for sound waves.

DSI Deep Sky Imager CCD camera.

DSLR Digital Single Lens Reflex camera. Sighting is done through the lens.

Dynamic Range Digital cameras with 8-bit Analog-to-Digital-Converters (ADC) have a dynamic range of 256 levels. Cameras with 16-bit ADCs have a dynamic range of 65,536 levels. Color cameras with 24, 32 or 48 have dynamic ranges of 256–4,096 levels.

Eclipsing Binary This is a star system made up of 2 or more stars that revolve around each other in the plane of the line of sight to Earth.

Emission Spectra Emission spectra are spectra with discrete emission lines corresponding to the electron energy released in elements present in the source.

Envisage Envisage is a program by Meade used for control of DSI CCD cameras and acquisition of images.

Epsilon Aurigae A famous bright long period (27.1 years) eclipsing binary star system.

eShel eShel is a high-resolution spectrograph based on an Echell design that allows high-resolution imaging of the whole visible spectrum at once by using higher order spectra.

EW Equivalent Width. The area under a curve representing the power of the curve referenced to one unit of Intensity.

Extinction This is the decreased brightness of stars when they are measured at lower elevations in the sky, due to particles in Earth's atmosphere scattering part of the light.

Extrinsic Variable This is a type of variable star that varies due to something external to the star, such as an eclipsing binary star system.

Extraterrestrial Magnitude (zero air mass) This is an extrapolation of measured magnitudes with air mass variation to estimate the brightness of starlight arriving at the top of our atmosphere.

FITS Flexible Image Transport System, a file format specification used in astronomy to convey both large bit array images and extended header information not possible with other formats like JPEG, etc.

First Order Spectrum Spectra are produced as higher levels on each side of the zero order spectrum. There are two first order spectra, one on each side of the zero order spectrum.

Flat Field Flat images that have a dark frame subtracted. These are used to create flat frames by dividing the image by the average pixel intensity ADU count.

Flux The Y-axis of a line profile is the intensity of the spectrum. This intensity is related to the number of photons detected at a specific wavelength. The flux is defined as # photons/cm^2/sec/nm.

Fourier Transform This is a mathematical theorem that enables any continuous function to be represented by a series of sine or cosine functions, which can be more readily analyzed.

FOV Field of View.

Flat Frame Flat frames are flat fields that have had all the pixels divided by the average pixel ADU count. The flat frame is then used to calibrate a spectrum image by dividing the flat frame into the spectrum image.

Fraunhofer Lines These are some 700 dark absorption lines in the spectrum of the Sun. These were first studied and named by Joseph von Fraunhofer in 1814.

Frequency For a cyclic phenomena, frequency is the rate of the cycle or cycles per second.

FWHM Full-Width-Half-Maximum.

Geocentric Time The time of an event based on the time at the center of the Earth.

Globular Clusters A large ball of stars located on the edges of the Milky Way galaxy.

Grism A grism is a wedge prism used in conjunction with a diffraction grating, usually in front of the grating to increase the resolution of the system.

Guiding Camera A CCD camera that is attached to the Guide Port of a spectrograph and used to keep the object of interest centered in the slit of the spectrograph.

Gunn James Edward Gunn, an astronomer who designed a digital camera to image galaxies and quasars.

Hale George Ellery Hale (29 June 1868–21 February 1938). He was famous for constructing the Yerkes 40″ reflector telescope north of Chicago on Williams Bay, 100″ Hooker reflector on Mount Wilson and the Hale 200″ reflector on Mount Palomar.

Heliocentric Time The time of an event based on the time at the center of the Sun. For astronomical events on the opposite side of the Sun relative to the Earth the maximum time correction is −8.3 min. For astronomical events beyond the Earth on the same side of the Sun as the Earth the maximum time correction is +8.3 min.

Heliocentric Wavelength Correction A correction of the Earth's motion relative to an astronomical object of interest. This is a wavelength correction that is added or subtracted to a measured spectral line's wavelength.

Heliocentric Julian Date (HJD) Correcting an earth-based observation for the light travel time to the Sun's center, in order to be able to precisely compare observations at different times of year.

Hooker John D, Hooker provided the financial backing for George Hale's 100″ telescope on Mount Wilson, California. The telescope bares his name, the Hooker Telescope.

Hot Pixels Pixels of a CCD or CMOS camera that have gone into saturation due to thermal electrons.

Hubble Edwin Powell Hubble (30 November 1889–28 September 1953), was the first person to determine the distance to the Andromeda Galaxy proving it was an island far distant from the Milky Way. He also determined that the Galaxies were receding and the Universe was expanding. This became what is known as "Hubble's Law."

Hydrogen Alpha The first Balmer series transition in hydrogen, from level 2 to level 3.

Hydrogen Balmer Lines These are prominent lines in the spectrum of hydrogen.

Image Processing The computer processing of an image to produce a suitable image with which to create a line profile.

Intrinsic Variable A type of variable star that varies due to something happening within or on the star. This can be a pulsation, like the Cepheids, or star spots.

Instrumental Magnitude The raw photometric magnitude corrected for the instrument used.

Interferometry Interferometry is a method of combining light from two or more telescopes to create an interference pattern that achieves the equivalent resolution of a telescope as large as the separation of the component telescopes.

Imaging Camera Some spectrographs use two CCD cameras, an imaging camera focused on the spectrum to image the spectrum and a guiding camera focused on the slit to allow tracking and keeping the object of interest in the slit.

Inverse-Square Law At a given solid angle the area increases inversely as the square of the distance. If the distance is doubled, the area is increased by four. At three times the distance the area is increased by nine. For a radiating body, the intensity per unit area is decreased four and nine times respectively.

ISIS Innovative Spectrographic Integrated Software. This is a spectral image processing software application used primarily with the Lisa spectrograph, but suitable for other spectrographs. It is a French freeware program written by Christian Buil.

IRIS IRIS is an image processing software application used for processing spectra. It is a French freeware program written by Christian Buil.

Julian Date (JD) JD is a sequential date, starting at noon on 01 January 4713 B.C., counting system in astronomy, useful for easy comparison of observations.

KH-11 Also know as the Kennan Satellite. A super secret reconnaissance spy satellite with code names Key Note and 101. This was a telescope very similar to the Hubble Space Telescope, but designed to look down toward the Earth rather than at the stars.

Kirchoff's Laws Three empirical rules for spectra analysis:

Linearity Linearity is the ability to fill a CCD pixel well with electrons in a linear way.

Lhires Littrow High Resolution Spectrograph.

Lhires III The Lhires III is a high resolution (2,400 l/mm grating) spectrograph.

Lhires Lite The Lhires Lite is a spectroscope (not a spectrograph) similar to the Lhires III with a 2,400 l/mm grating, but designed for solar observing without a telescope.

Linear Straight as a straight line.

Linearity A measure of how linear or straight something is, such, as a function to calibrate wavelength or the change of ADU counts versus exposure time.

LISA Long Imaging-slit Spectrograph for Astronomy. The LISA is a high-precision low to medium-resolution spectrograph.

Littrow Joseph Johnann von Littrow (1781–1840). In 1833 he created what is known as the Littrow projection. A Littrow spectrograph is one that has a light path that folds back on itself.

Luminosity Classes Luminosity Classes are distinctions made among types of stars based on surface gravity, arising from different interior evolutionary states; includes supergiant, giant, dwarf and white dwarf stars (the Sun is a dwarf star).

Lyman Series The Lyman Series is a sequence of spectral lines formed when electrons in hydrogen atoms transition between level 1 and higher levels – ultraviolet part of the spectrum.

MaxIm DL A commercial imaging program that allows camera control of many different CCD cameras and image processing.

Measure Lines Vertical lines in a spectrum line profile window that delimit an area of interest to be measured.

Messier Marathon Over a single night, usually in March of each year, astronomers attempt to observe all 110 Messier objects. The objects consist of galaxies, nebulae and star clusters that Charlie Messier observed and cataloged in 1771.

Milky Way The galaxy in which we live. It was once thought to comprise the entire universe.

MK System A method of classifying stars in terms of spectral type and luminosity class, based on their spectrum appearance; the Sun is a G2 dwarf in this system; epsilon Aurigae resembles an F0 supergiant. Devised by W.W. Morgan, and P.C. Keenan.

Modified Julian Date (MJD) Modified Julian Day (MJD) – number of days that have elapsed since midnight at the beginning of a defined, cardinal number Julian Date like 2,400,000.5.

Mount Palomar A 5,700¢ high mountain located above Pasadena, California that hosts the famous 100″ Hooker telescope along with a variety of other astronomical instruments.

Mount Wilson A 6,100¢ high mountain located North of San Diego, California that hosts the famous 200″ Hale telescope.

Nanometer (nm) Nanometer, a measure of wavelength. 1 nm = 10 Ångströms.

Nebulae These are clouds of gas and dust between stars and the results of supernovae.

NIST National Institute of Standards and Technology.

Normalization To be able to determine an equivalent width of a line the spectrum line profile must be normalized. To normalize a line, first an average intensity, ADU count, in the area of the line of interest is determine. Next, that number is divided into the continuum. This produces a continuum line profile centered on the value of 1.00.

OEM Original Equipment Manufacturer.

Orion Image Suite A commercial imaging software program included with the Orion StarShoot cameras that allows control of the camera and imaging processing.

Partial Julian Date When plotting data, for clarity a "partial" Julian Date (PJD) which is the Julian Date minus a fixed quantity, e.g., PJD = JD − 2,450,000 may be used.

Parallax The apparent shift of an object when viewed from different directions.

Parsec (Parallax Second) Parsec is a natural unit of triangulation, relating earth's orbit (AU) as a baseline, and a 1 arc second angular deviation, equivalent to 3.3 light years or 31 trillion kilometers.

Peltier Jean Charles Athanase Peltier (22 February 1785–27 October 1845), a French physicist who in 1834 discovered the Peltier Effect.

Peltier Effect A junction that when a DC voltage is applied one side of the junction gets hot and the other side gets cold. Also known as the thermoelectric effect.

PHD PHD is a freeware program that allows a star to be selected and then controls the telescope to track that star. The program has been used with good results for high-resolution guiding.

Photometry Astronomical Photometry is the measurement of the flux or intensity of a star's electromagnetic radiation.

Pier A permanent mount for a telescope. Usually set in concrete and isolated from the surrounding area.

Pixel Picture element.

Pixel Map A map of the pixels on a CCD chip that shows the intensity ADU counts for each pixel.

Planck's Constant Planck's constant is the physical constant of value $h = 6.57 \times 10 - 27$ erg-sec. This is the quantum of action in quantum mechanics.

Planck Curve The spectrum of a star produces a Planck curve that obeys Wein's Law. The peak intensity of the Planck Curve indicates the temperature of the star.

Planck's Law Planck's Law describes the relation of a black body's temperature to the intensity of the radiation.

PMT Photomultiplier Tube. A very sensitive light detector used for photometry.

Polynomial An algebraic expression with order higher than 2. $Y = a0 + a1X + a2X2 + a3X3 + \ldots anXn$

Prime Focus Spectrograph A diffraction grating spectrograph used at prime focus on the 200″ Hale telescope. It was the first Earth-based astronomical spectrograph to use a digital CCD camera.

Prism A glass optical element in triangular shape.

Quasars These are star like objects that have enormous red shifts indicating they are billions of light years away. The mystery is their size is on the order of the size of our solar system, but their energy output exceeds that of a whole galaxy.

Radial Velocity Radial Velocity is a spectroscopic measurement of the Doppler shift of light due to motion of the source toward or away from the observer; same principle as is used in traffic radar measurement.

Radiometry Radiometry is a means of measuring characteristics of electromagnetic radiation, typically the power or intensity (flux) distribution of photons.

Rainbow Optics A low cost low-resolution transmission grating spectrograph with a 200 l/mm grating that fits as a filter cell.

Raw Flat Image The image of a flat field (uniformly white light exposure) before dark frame subtraction.

Raw Magnitude The RAW magnitude is determined by $-2.5 \log (I)$, where I is a number (typically ADU counts) representing the measured intensity of a star.

Reduced Julian Date (RJD) Reduced Julian Day (RJD) counts days from nearly the same defined day as the MJD, but lacks the additional offset of 12 h used in MJD.

Resolution The ability to see fine detail.

Rotational Broadening Rotational Broadening is a spectroscopic method for determining rotation of a source based on Doppler widening of spectral features in proportion to rotation speed and orientation.

RSpec Realtime **Spec**troscopy software. This is an easy to learn, powerful and intuitive commercial spectrum processing software program. Tom Field developed this program.

Saturation Saturation is the point where the pixel ADU count cannot be exceeded, 65,536 counts for a 16-bit camera and 256 counts for an 8/24-bit camera.

Schmidt Maarten Schmidt. (28 December 1929) Pursued the mystery of Quasars.

Secchi Angelo Secchi. In 1866 discovered the first Be star, gamma Cassiopeiae

Shelyak The star beta Lyrae and the name of the company that makes several high-quality spectrographs. http://www.shelyak.com

Slit A small rectangular opening in the light path before a diffraction grating or prism. Typically the slit is 15 to 100 mm wide.

Slitless Using a spectrograph without a slit. A star image or small fiber optic provides the equivalent of the slit.

Slit Plate A plate used on the ALPY 600 Basic Module that contains several selectable slits.

Sodium D Lines The sodium D lines are two prominent closely spaced lines in the spectrum of sodium. D1 = 5,889.910 Å and D22 = 5,895.924 Å. These lines give the spectrum of sodium its bright yellow color.

Sono Tube A heavy cardboard tube of various diameters used in construction as a form for concrete columns. The Sono tube is ideal for making telescope piers.

Spectrograph When a device capable of taking spectra of astronomical objects is connected to a telescope and digital camera, the device is called a spectrograph.

Spectrohelioscope The spectrohelioscope or spectroheliograph is special spectrograph designed specifically for solar observations.

Spectrometer The term spectrometer is interchangeable with a spectrograph.

Spectrophotometer A photometer specifically designed to measure intensity (flux) of photons at specific wavelengths. Its main use is with chemistry and not astronomy.

Spectroscope A device used to display the visible spectrum of light visually. A prism or diffraction grating may be used.

Spectra The plural of spectrum.

Spectroscopy the method of dispersing light into component colors (e.g. rainbows); analysis of the spectrum can yield information about the temperature, motion and chemical composition of the source of light.

Spectrum The range of electromagnetic spectrum from just above zero hertz at the low-frequency, long wavelengths, and low-energy end to high-energy gamma rays at the high end of the spectrum, short wavelengths.

Spectrum Processing The processing of a spectrum images to produce a calibrated line profile.

Spline A software tool used to create a smooth curve.

SpcAudACE A software program for processing spectra. This is a French freeware program written by Benedict Maugis.

Star Analyser A low cost low-resolution blazed transmission grating spectrograph with a 100 l/mm grating that fits as a filter cell.

Stark Effect The Stark Effect is a type of spectroscopic line broadening due to strong electric fields near the source of the light being measured. It is also a pressure broadening effect.

Stokes Parameters Stokes Parameters as used in polarimetry, measurements of the amplitude of perpendicular components of the electromagnetic field mutually perpendicular to the direction of light travel.

Telluric Lines Telluric lines are spectral absorption lines due to photon absorption of oxygen, water and carbon dioxide in the Earth's atmosphere. Also known as atmospheric lines.

Thermoelectric Effect Also known as the Peltier Effect. The effect of using a DC current to cool a junction. Used in thermoelectric coolers.

ToUcam An inexpensive color web camera. This was slightly modified and used in the early days of astrophotography.

Vega Also known as alpha Lyrae. α Lyrae is a bright type A star with a spectrum that is easy to analyze.

V/R The ratio of a violet, short wavelength emission line to a red, long wavelength emission line.

VSpec Visual Spec. A software application used for processing spectra. This is French freeware written by Valerie Desnoux.

Wavelength Wavelength is the distance between two corresponding points of a varying function and is the inverse of frequency, cycles per unit of time. Wavelength is denoted by the Greek letter lambda λ. The letter c is the velocity of the wave.

$\lambda = c/\text{Frequency}$

Wien's Displacement Law A law that states that the wavelength distribution of thermal radiation from a black body at any temperature has the same shape as the distribution of any other temperature except each wavelength peak is display with the higher temperature peaks being displaced toward shorter wavelengths.

Wiffpic WF/PC or Wide-Field/Planetary Camera, a CCD camera used on the Hubble Space Telescope.

Wolf-Rayet Stars Stars that are very hot and massive (25 solar masses or more) at the end of their evolution that unlike most other stars have spectra with emission lines rather than absorption lines.

WR Stars Wolf-Rayet Stars

Yerkes Charles T. Yerkes financed George Hale's 40″ refractor telescope, known as the Yerkes Telescope, North of Chicago. It is the largest refractor telescope in the world.

Zeeman Effect The Zeeman Effect is a type of spectroscopic line broadening due to strong magnetic fields near the source of the light being measured.

Zero Order This spectrum is at 0 Å and is essentially the star's image.

Zenith This is the point directly over an observer, the shortest distance to outside the Earth's atmosphere. Air mass = 1.0.

About the Author

Jeffrey L. Hopkins has been interested in astronomy most of his life. His first telescope was a 2.4" Unitron refractor, which he still has. Back in the mid-1950s, the telescope was paid for over many months with funds from a newspaper route. His first serious contributions to astronomy began in the early 1980s. After a non-productive effort to build and use a phase-switching interferometer radio telescope, he designed and built a UBV photon counting photometer for use on his C-8 telescope. Astronomical photometry occupied most of his observing time until mid-2006 when he began doing high-resolution spectroscopy on the mysterious star system epsilon Aurigae.

Hopkins has a BS from Syracuse University with graduate work in physics at the University of Wyoming, University of Arizona and Arizona State University. He was a Radar Systems Engineer for General Electric Company for 13 years and a Senior Engineer for Motorola for 10 years. He has written or co-authored over four dozen professional astronomical papers as well as five other astronomy-related books.

In the spring of 2011, Hopkins had a minor planet named after him for his astronomical work on the mysterious long period eclipsing binary star system epsilon Aurigae. The minor planet (asteroid) number is 187283. At the August 2012 meeting of the Astronomical Society of the Pacific, he was presented the prestigious *Amateur Achievement Award*. He is originally from upstate New York, but now resides in Phoenix, Arizona. During the clear cooler nights, he spends his time in one of his two backyard observatories.

phxjeff@hposoft.com
http://www.hposoft.com/Astro/astro.html

Index

A

Absolute calibration, 32, 269
Absolute magnitude, 34, 36, 37, 93–98, 270
Absorption spectra, 6, 8
ADC. *See* Analog to Digital Converter (ADC)
Adjusting the Lhires III, 157
ADU. *See* Analog to Digital Unit (ADU)
Advanced options, 179
AIP4WIN, 269
Air mass, 32, 100, 173, 224, 269, 272, 278
Aldebaran, 94–95, 269
All Electronics, 104, 165
Alnilam, 243
Alpha Lyrae, 93, 269, 277
ALPY 600, 32, 46, 55, 57, 103, 108, 125–150, 276
ALT/AZ, 42, 269
Amateur spectroscopy, 235, 239
Analog to Digital Converter (ADC), 53, 68, 70, 151, 269, 271
Analog to Digital Unit (ADU), 17, 19, 22, 32, 50, 62, 64–66, 68–71, 73, 76, 81, 82, 90, 99, 100, 116, 162, 172–174, 186, 190, 212, 221, 269, 270, 272, 274–276
Angelo Secchi, 248, 276
Ångströms (Å), 4, 269
Apparent magnitude, 34–36, 93–99
Appearance menu, 194–195, 202

Arc second, 38, 270, 275
ASCII, 121, 122, 270
Astronomical spectroscopy theory, 8–32
ATIK 314L+, 52–54
Atmospheric lines, 270, 277
Atomic hydrogen, 5–6
AU, 88, 270, 275
AutoStar, 60, 171, 270
A0V star spectrum, 13

B

Background, 3, 7, 55, 72–73, 77, 91, 116, 145, 163, 180, 183, 184, 187–189, 197, 201, 202, 271
Balmer Series, 4, 270, 273
Barycenter, 166, 192, 193, 205, 206, 270
Basic module, 57, 125–132, 133, 135–138, 140–143, 149, 150, 276
Bellatrix, 98, 270
Be Stars, 245, 248–249, 270, 276
Beta Lyrae, 249, 270, 276
Betelgeuse, 13, 34, 86, 96, 162, 181, 238–240, 269, 270
Binning, 71, 73, 163, 173, 179, 185, 239, 240, 269, 270
Black body, 10, 270, 276, 278
Blazed grating, 75–77, 270
Bleau, Willem Janszoon, 245
Brackett Series, 270

Index

Bubble wavelength, 204
B0V star, 14
B&W Tek, 102

C
Calibrate, 17, 26–28, 32, 45, 61, 68, 70, 81, 92, 97, 99, 103, 116, 117, 142, 145–147, 159, 163, 164, 166, 172–175, 179, 180, 185, 200, 203, 204, 206, 208–210, 212, 213, 218, 219, 221, 229, 272, 274, 277
Calibration
  menu, 204
  module, 57, 125, 126, 138–139, 147
Capella, 94, 95, 235
Cassiopeiae, 248, 249, 276
CCD. See Charge coupled device (CCD)
CD ROM, 102, 127
Centroid, 270
Cepheid, 36–38, 270, 273
CFL, 102, 111, 123, 270
Charge-bubble device, 41, 270
Charge-coupled device (CCD), 41–43
  cameras, 41–43, 50–54, 57, 58, 62, 64, 80, 82, 83, 85, 86, 90, 93, 123, 126–130, 132, 134, 141, 143–145, 151, 153, 169, 171, 176, 186, 212, 263, 269–274, 276, 278
  chip pixels, 46
  spectra, 50
Chemical composition, 16–17, 277
CMOS, 49–52, 68, 81, 176, 186, 271, 273
Collisional broadening, 9, 10, 271
Continuous spectrum, 3, 7–9, 62, 70, 130, 132, 141, 271
Converging beam, 77, 82, 88, 89
Correction procedure, 212–219
Czemy-Turner design, 102

D
Dark frame, 63, 64, 68, 69, 72, 91, 103, 114, 171, 173, 180, 181, 265, 271, 272, 276
Data acquisition, 271
Data reduction, 172–173, 271
Decimal date, 253, 255–256, 271
Delete line or feature, 201–202
Delete range, 199–201
Delta Orionis, 241–242
Delta Scorpii, 244, 245, 248
Deneb, 95
Deselect legend visible, 198

Diffraction grating, 42, 55, 61, 75–77, 79, 108–111, 270–272, 276, 277
Digital camera, 17, 41, 49, 50, 79, 99, 126, 140, 151, 270–272, 277
Digital Single Lens Reflex (DSLR), 46, 49, 50, 77, 79–83, 88, 89, 91, 151, 271
Dispersion, 23, 27, 28, 30, 79, 83, 85, 163, 204, 206, 207, 229, 271
Do-It-Yourself (DIY)
  Excel, 171, 174–175
  spectroscope, 101–123
Doppler
  broadening, 9, 10, 18
  effect, 18, 271
DSI, 51–54, 85, 86, 144, 145, 171, 265, 271
DSLR. See Digital Single Lens Reflex(DSLR)
Dynamic range(s), 49, 50, 62, 82, 176, 271

E
Earth's radial velocity,31
eBay, 56, 101, 103, 123
Eclipsing binary, 8, 235, 271, 272
Edit menu, 199
Electromagnetic spectrum, 2–6, 55, 277
Emission spectrum, 3, 7
Envisage, 145, 264, 271
Epsilon Aurigae, 13, 220, 235–238, 248, 255, 271, 274
Epsilon Orionis, 243–244
Equivalent width (EW), 18, 20–21, 45, 176, 193, 203, 221–222, 271, 275
eShel, 55, 59–60, 108, 151, 271
Excel, 22–26, 50, 117, 171, 174–175
Exposure, 42, 46, 50, 60, 62–64, 68, 70, 82, 86, 87, 89–91, 99, 100, 103, 104, 115, 126, 142, 145, 148, 159–162, 166, 168, 169, 173, 174, 180, 181, 185, 186, 235, 250, 263–265, 274, 276
Extinction, 32, 100, 173, 271
Extraterrestrial magnitude, 172, 173, 271, 272
Extrinsic variable, 33, 272
Eyepiece,77, 80, 82–84, 89, 106, 113, 143

F
Field of view (FOV), 272
Finding the object, 88–89
First order spectrum, 272
FITS. See Flexible image transport system (FITS)
Flat field, 272
Flat frames, 68–72, 91, 272

Index

Flexible image transport system (FITS), 176, 179, 263–265, 272
Flux, 26, 272
Focusing, 88, 89, 106, 108–111, 130, 132, 133, 135, 154, 158–164, 168
Fourier transform, 272
Fraunhofer, 8, 9, 272
Frequency, 2, 4, 165, 271, 272, 277, 278
Full width half maximum (FWHM), 20, 193, 272
F0V Star, 14
FWHM. *See* Full width half maximum (FWHM)

G
Gamma Cassiopeiae, 248, 276
Geocentric time, 272
Globular clusters, 34, 272
Graph, 17, 22, 24, 26, 116, 168, 174, 194, 197–199, 201, 204, 209
Gratings, 29, 30, 42, 55–59, 61, 62, 75–77, 79, 83, 84, 90, 102, 103, 108–111, 125, 153–156, 161, 163, 166, 168, 240, 242, 243, 264, 270–272, 274–277
Grism, 57, 80–81, 125, 127, 272
Guiding camera, 133, 134, 136, 158–159, 162, 169, 272, 273
Guiding module, 57, 125–127, 129, 133–137, 140, 143–147, 149
Gunn, James, 42, 269, 272
G0V Star, 14, 16
G5V Star, 15

H
Hale, 37, 38, 40–42, 250, 269, 272, 275, 276, 278
Heliocentric
 calibration, 222
 correction, 31, 32, 176, 222, 230–231, 258
 time, 272
 wavelength correction, 30–32, 222, 231, 273
Heliocentric Julian Date (HJD), 253, 258, 273
High-resolution, 20, 23, 28, 30, 45, 46, 49, 51, 58, 60, 68, 70, 72, 73, 151, 152, 158, 162–168, 176, 179, 181, 184, 186, 188, 190, 205, 208, 212, 220, 227, 239, 242, 243, 248, 250, 271, 274, 275
Histogram, 162, 182-
 sliders, 183
 tool, 182, 183
HJD. *See* Heliocentric Julian Date (HJD)
Hooker, 38, 39, 272, 273, 275

Horizontal binning, 73, 173, 185, 190
Hot pixels, 63, 64, 68, 273
Hubble, 38–41, 273, 274, 278
Hydrogen
 Balmer lines, 12, 42, 92, 93, 99, 210, 228, 234, 267, 273
 Balmer spectrum, 6
 series, 5
 transition series, 5

I
Image
 camera orientation, 158
 processing, 17, 65, 66, 91, 171, 173, 176, 181, 269, 270, 273, 274
 rotation, 72, 91, 163
Imaging camera, 72, 113, 157–160, 168, 169, 273
Instrument
 response, 92, 193, 203
 spectral response, 198
Instrumental magnitude, 172, 273
Interferometry, 273
Intrinsic variable, 36, 270, 273
Inverse-square law, 34–36, 272, 273
IRIS, 273
ISIS, 273

J
Julian, 231, 253, 256–261, 273–276

K
Kennan, 41, 274
Key Note, 41, 274
KH-11, 41, 274
Kirchoff's Laws, 3, 274
K0V star spectrum, 15, 16

L
Labels menu, 195
Laser pointer, 129, 155–156, 168
Lhires III, 30, 46, 52, 54, 55, 58–59, 61, 169
Line(s)
 broadening, 9–10, 12, 277, 278
 profile, 97, 99, 103
Linear, 22, 23, 29, 62–63, 80, 90, 99, 102, 108, 117, 163, 174, 179, 204, 206, 208, 212, 214, 221, 227–229, 274
Linearity, 62–63, 90, 163, 186, 274
LISA, 46, 55, 57, 58, 151, 273, 274

Littrow, 55, 152, 161, 274
Loading the spectra, 181–182
Locking ring, 79–80, 84, 85, 90, 129, 132, 133, 136
Low-resolution, 20, 23, 28, 30, 32, 45, 46, 49, 55, 57, 60, 68, 69, 72–73, 75, 85–99, 149, 163, 179, 181, 187, 188, 190, 204, 206, 212, 220, 227, 235, 240, 242, 245, 248, 250, 276, 277
Luminosity classes, 12–13, 274
LX90, 47–49, 86, 135
LX200, 46, 157, 240, 243, 265
Lyman Series, 274

M
M05, 15
Math, 202–203, 217, 219, 221
MaxIm DL, 171
Meade DSI Pro series, 51
Measure
    lines, 166, 192–193, 201, 202, 205, 206, 213, 215, 221
    menu, 191–192
Measurement options, 192
Messier Marathon, 274
Micrometer (μm), 153–156, 158, 163, 168
    calibration, 154–156
Milky Way, 33, 34, 38
MJD. *See* Modified Julian Date (MJD)
MK of spectral type, 12
MK system, 11
Modified Julian Date (MJD), 253, 257–, 274
Mount Palomar, 40–43
Mount Wilson, 38–40, 275
Multiple line calibration, 28, 29, 227

N
Nanometer (nm), 4, 116, 117, 121, 180, 275
National Institute of Standards and Technology (NIST), 119, 275
Nebulae, 34, 37, 275
Neon, 28, 73, 132, 138, 141–143, 147–149, 154, 155, 158–166, 168
    lamp ring wiring, 165
NIST. *See* National Institute of Standards and Technology (NIST)
Non-converging, 77, 79, 82, 83, 88
Non-linear, 22–29, 80, 145, 146, 163, 179, 204, 208–209, 227–229
Normalization, 18–19, 221, 275

O
Original Equipment Manufacturer (OEM), 102, 275
Omega Orionis, 249
One point calibration, 28
Optical bench, 102, 108–112
Options, 118–120, 126, 151, 154, 178–180, 185–187, 189, 192, 194, 198, 202, 219
Orion Image Suite, 275
Orion StarShoot, 51–54, 129, 130, 144, 145, 171, 265, 275

P
Parallax, 33–34, 275
Parsec (Parallax Second), 275
Partial Julian Date (PJD), 275
P Cygni, 245–247
Peltier, 43, 275
PHD, 275
Photometry, 68, 172, 246, 250, 258, 275
Pier, 46–49, 275
Pixel
    count, 41, 67, 176
    map, 17, 64–70, 185–186, 192, 275
PJD. *See* Partial Julian Date (PJD)
Planck curve, 12, 275
Planck's constant, 4, 275
Planck's Law, 276
PMT, 172, 276
Points visible, 196
Pollmann, Ernst, 246
Polynomial, 23–26, 28, 117, 276
Prime Focus System, 42, 276
Prism, 1, 2, 9, 39, 42, 55, 57, 80, 81, 125, 276, 277
Processing, 17, 22, 23, 45, 61, 65–68, 91–99, 145, 146, 163–168, 171–231, 273, 277
Program options, 178, 179
Pseudo spectra, 220

Q
Quasars, 42, 227, 234, 250, 276

R
Radial velocities, 18, 30, 31, 39, 231, 245, 276
Radiometry, 276
Rainbow optics, 55, 75, 76, 78, 276
Raw flat image, 276
Raw magnitude, 172, 173, 271, 276
Reference menu, 198–199, 202

Regular, 76, 96, 173, 176, 239
Resolution, 20, 23, 28, 30, 46, 49, 50, 55–58, 60, 68, 70, 72, 73, 75, 77, 92–99, 102, 125–150, 162–168, 176, 179, 181, 190, 206, 212, 220, 227, 240–243, 250, 276
Resolution *versus* Dispersion, 30
Response
 calibration, 193, 202, 212–219
 curves, 54
Rigel, 90–91
RJD. *See* Reduced Julian Date (RJD)
RK0V Star, 16
Rotate and slant, 186–187
Rotational broadening, 9, 10, 276
RSpec, 13, 22, 27, 64, 66, 73, 83, 85, 103, 142, 146, 166, 171, 175–222, 227, 276

S
Saturation, 62–64, 70, 90, 186, 272, 276
Schmidt, Maarten, 42, 276
Science-Surplus.com, 102
ScopeStuff, 51, 82, 84, 134
Shelyak, 56–60, 80, 108, 126, 129, 130, 138, 139, 149, 151, 152, 154, 249, 270, 276
4-Shooter, 41, 42
Side plates, 159, 160
Sidereal time, 253, 255
Single point, 28
Slant, 186–187, 270
Slit, 1, 55, 57, 58, 71, 77, 99, 102, 104, 108–111, 125–129, 131–133, 135, 136, 141–143, 145, 147, 149, 153, 157–159, 161–164, 169, 181, 188, 276
Slitless, 55, 57, 79, 104, 128, 129, 143, 147, 181, 187, 276
Slit plate, 126–129, 131, 133, 135, 141, 143, 149, 276
SMA, 102, 104, 105, 108
Sodium D lines, 2, 147, 148, 155, 166, 167, 206, 211, 220, 233, 238, 267, 277
Solar, 9, 35, 37, 42, 92–93, 100, 107, 112, 234, 243, 247, 256, 274, 276–278
Sono tube, 47, 277
Sony, 52, 102, 103
SpcAudACE, 277
Spectra (l)
 order, 29, 60, 61, 89, 163
 type, 11–13, 274
Spectrograph, 29, 33, 39, 42, 46, 55–60, 72, 75–79, 92, 101, 104, 105, 108, 109, 112, 113, 126, 139, 140, 152–164, 166, 168, 169, 174, 181, 187, 212, 241, 250, 264, 276
Spectroheliograph, 277

Spectrohelioscope, 277
Spectrometer, 76, 101–112, 114, 122, 123, 277
Spectrophotometer, 32, 277
Spectroscopy, 1, 45, 75, 101, 125, 151, 171, 233
Spectrum
 focusing, 132, 159–162
 window, 102, 108, 109, 111, 130, 178–181, 183, 185, 212, 225
Spline, 214, 216, 277
SS Leporis, 249
Star analyser, 30, 32, 46–48, 51, 55–57, 60, 61, 73, 75–100, 103, 128, 143, 176, 181, 188, 204, 207, 240, 241, 248, 250, 277
Stark effect, 9, 11, 277
StarShoot, 51–54, 129, 130, 144, 145, 171, 265, 275
Stellar spectra classification, 11
Stokes parameters, 277
Sun, 7–10, 12, 13, 30, 36, 92, 95–100, 103, 158, 234, 245, 254, 255, 258, 270, 272–274
Synthesize, 219–220

T
Taking spectra, 60–73, 147, 277
Tangential velocity, 31
Telescope, 33, 46, 75, 104, 125, 152, 171, 233
Telluric, 28, 100, 164, 227, 270, 277
Thermoelectric, 42, 43, 275, 277
Tilt, 72, 110, 145, 186, 254
Tools, 105, 116–120, 178, 182, 183, 192, 277
ToUcam, 84, 85, 180, 277
Trendline, 25
Trigonometry, 33, 34

U
Unfocused, 161
USB, 102

V
Vega, 32, 90, 93, 145, 169, 223, 269, 277
V/R, 22, 278
VSpec, 22, 32, 67, 103, 171, 175, 179, 222–230, 278

W
Wavelength, 2, 45, 75, 101, 126, 153, 173,
Wavelength correction, 30–32, 222, 231, 273
WG05 star, 16
Wien, 11, 12, 278

Wiffpic (WF/PC), 41, 278
Wolf-Rayet (WR) stars, 247–248, 278
Woodland Hills, 80, 126

Y
Yerkes, 37–38, 272, 278

Z
Zeeman effect, 10, 278
Zenith, 39, 41, 46, 100, 173, 269, 278
Zero order, 28, 29, 60–62, 76, 79, 83, 85, 86, 88, 89, 91, 163, 204, 206, 207, 272, 278
Zeta Tauri, 249

Printed in Great Britain
by Amazon